Angular 4

Pocket Primer

Angular 4
Pocket Primer

Oswald Campesato

Mercury Learning and Information
Dulles, Virginia
Boston, Massachusetts
New Delhi

Publisher: David Pallai

MERCURY LEARNING AND INFORMATION
22841 Quicksilver Drive
Dulles, VA 20166
info@merclearning.com
www.merclearning.com
(800) 232-0223

O. Campesato. *Angular 4 Pocket Primer.*
ISBN: 978-1-68392-035-9

Library of Congress Control Number: 2017934714
17181932   Printed in the United States of America on acid-free paper.

Our titles are available for adoption, license, or bulk purchase by institutions, corporations, etc. For additional information, please contact the Customer Service Dept. at (800) 232-0223(toll free).Digital versions of our titles are available at: www.authorcloudware.com and other electronic vendors. *Companion files are available from the publisher by writing to info@merclearning.com.*

I'd like to dedicate this book to my parents –
may this bring joy and happiness into their lives.

CONTENTS

PREFACE

What is the Primary Value Proposition for This Book?

This book endeavors to cover basic features of Angular 4, along with a variety of code samples (based on a production release of Angular 4) that will fit in a 300-page book. The primary goal of the code samples is to show you how to accomplish various tasks, such as displaying and updating a list of users, saving information to a data store, working with `Observables`, developing mobile applications, and so forth. If you are interested in reading Angular code samples that "get stuff done," even though some details are omitted, this book might be a reasonably good fit for you.

If you are undecided about the style of this book, or if you feel that a "pocket primer" ought to have a different format (and opinions do vary on this point), then perhaps a different book would be more suitable for your needs.

Is Knowledge of Angular Required for This Book?

No prior knowledge of Angular is required in order to read this book. You do need an understanding of HTML (how to use basic elements) and some knowledge of JavaScript. Chapter 6 contains Node-related code samples, which does require some background knowledge to understand the material. Chapter 7 contains code samples that combine Angular with Redux and GraphQL, which are more advanced topics. However, these two chapters can be omitted if you want to focus on "core" concepts in Angular. If you encounter concepts that are unfamiliar, in many cases you can learn them by reading one of the many online tutorials that explain those concepts.

Please keep in mind that there is always a trade-off between the depth and breadth of the explanation of the technical details of the code samples, as well as the type (and number) of samples to include in a short book. This book is inclined toward breadth rather than depth vis-à-vis the code samples. Consequently, some topics (such as testing) have been omitted, and other topics (ES6 and TypeScript) are barely covered, on an "as-needed" basis at best. As a result, sometimes you need to look elsewhere for more detailed information about the technical "bits" in the code samples, such as the ES6 methods map() and filter(), or an in-depth explanation of Observables and Promises.

If you want to be sure that you can grasp the material in this book, glance through some of the code samples to get an idea of how much is familiar to you and how much is new for you.

The Target Audience

This book is intended to reach an international audience of readers with highly diverse backgrounds in various age groups. While many readers know how to read English, their native spoken language is not English (which could be their second, third, or even fourth language). Consequently, this book uses standard English rather than colloquial expressions that might be confusing to those readers. As you know, many people learn by different types of imitation, which includes reading, writing, or hearing new material (yes, some basic videos are also available). This book takes these points into consideration in order to provide a comfortable and meaningful learning experience for the intended readers.

Getting the Most from This Book

Some programmers learn well from prose, others learn well from sample code (and lots of it), which means that there's no single style that can be used for everyone.

Moreover, some programmers want to run the code first, see what it does, and then return to the code to delve into the details (and others use the opposite approach).

Consequently, there are various types of code samples in this book: some are short, some are long, and other code samples "build" from earlier code samples.

How Was the Code for This Book Tested?

The code samples in this book have been tested in a Google Chrome browser (version 53) on a Macbook Pro with OS X 10.16. Unless otherwise noted, no special browser-specific features were used, which means that the code samples ought to work in Chrome on other platforms, and also in other modern browsers.

Another point to keep in mind is that all references to "Web Inspector" refer to the Web Inspector in Chrome, which differs from the Web Inspector in Safari. If you are using a different (but still modern) browser or an early version of Chrome, you might need to check online for the sequence of keystrokes that you need to follow to launch and view the Web Inspector. Navigate to this link for additional useful information:

http://benalman.com/projects/javascript-debug-console-log/

Why are the Screenshots in Black and White?

The black and white images are less costly than the original color images, and therefore their inclusion means that this book is available at a lower cost. However, the color images are available on the companion disc, along with supplemental code samples that render in color when you launch them in a browser.

Why Does This Book Have 300 Pages Instead of 500 Pages?

This book is part of a *Pocket Primer* series whose books are usually between 200 and 300 pages. Second, the target audience consists of readers ranging from beginners to intermediate in terms of their knowledge of HTML and JavaScript. During the preparation of this book, every effort has been made to accommodate those readers so that they will be adequately prepared to explore more advanced features of SVG during their self study.

Does this Book contain Production-Level Code Samples?

The code samples in this book do use a production release of Angular to illustrate various features of Angular. Clarity has higher priority than writing more compact code that is more difficult to understand (and possibly more prone to bugs), especially in a short book. If you decide to use any of the code in this book in a production website, you ought to subject that code to the same rigorous analysis as any other Web applications.

Other Related Books by the Author

1) HTML5 Canvas and CSS3:
http://www.amazon.com/HTML5-Canvas-CSS3-Graphics-Primer/dp/1936420341
2) jQuery, HTML5, and CSS3:
http://www.amazon.com/jQuery-HTML5-Mobile-Desktop-Devices/dp/1938549031
3) HTML5 Pocket Primer:
http://www.amazon.com/HTML5-Pocket-Primer-Oswald-Campesato/dp/1938549104
4) jQuery Pocket Primer:
http://www.amazon.com/dp/1938549147

The following open source projects contain code samples that sometimes supplement the material in various chapters of this book:

https://code.google.com/p/css3-graphics/

https://code.google.com/p/d3-graphics/

https://code.google.com/p/html5-canvas-graphics/

https://code.google.com/p/jquery-css3-graphics/

https://code.google.com/p/raphael-graphics/

https://code.google.com/p/svg-filters-graphics/

https://code.google.com/p/svg-graphics/

CHAPTER 1

QUICK INTRODUCTION TO ANGULAR

This chapter provides a fast introduction to developing Web applications in Angular. After covering some of the high-level aspects of Angular, you can quickly grasp many of the code samples that are discussed in later chapters (but some code samples are more extensive). At the same time, keep in mind that you need to invest additional time and effort to acquire a deeper understanding of Angular.

NOTE *The Angular code samples in this book are based on the Angular production code that was released in March 2017.*

Another important consideration is your learning style: you might prefer to read the details regarding the "scaffolding" for Angular applications before you delve into the first code sample. However, it's perfectly acceptable to skim the introductory portion of this chapter, then quickly "get into the weeds" with the first Angular sample code in this chapter, and later review the initial portion again.

The first part of this chapter discusses the design goals of Angular and some new features, such as components, modules, and one-way data binding. The second part of this chapter discusses the Angular command-line interface (CLI), which is a command-line tool for generating Angular applications.

NOTE *The Angular projects in this book are based on version `1.0.0` of `ng` (which is the Angular CLI) for creating Angular 4 applications.*

There are several points to keep in mind before you read this chapter. First, this book provides many examples of Angular applications, so the details about Angular concepts, design goals, and architecture are "lighter" than what you might find in 500-page books. However, you can find useful information in online articles by Victor Savkin (and other people).

Second, you can create Angular applications using ECMA5, ES6, and TypeScript. However, the main recommendation is to develop Angular applications using ES6 or TypeScript; in fact, all the code samples in this book use TypeScript, which makes the transpilation process (performed "behind the scenes") straightforward. Third, no previous experience with Angular is required, but some knowledge can obviously be helpful.

NOTE *When you copy a project directory from the companion disc, if the node_modules directory is not present, then copy the top-level node_modules directory that has been soft-linked inside that project directory (which is true for most of the sample applications).*

Supported Versions of Angular: How It Works

In case you didn't already know, the Angular landscape is moving fast, starting from the production release of Angular 4 in March 2017 and extending through Angular 7, which is scheduled for September 2018.

Each upcoming release (starting with Angular 6 in 2018) will probably support only two earlier versions of Angular. For example, Angular 6 will support Angular 4 and Angular 5, whereas Angular 7 will support Angular 6 and Angular 5 (but not Angular 4).

In addition, there is a sort of "duality" in the naming convention for Angular: upcoming Angular releases specify a version number, but Angular *without* a version number is the official name for Angular 2 and beyond (whereas AngularJS is the official name for version 1.x).

Another important goal of the Angular core team during this development process is to minimize breaking changes and ensure that such changes will have minimal impact with respect to upgrading to new versions of Angular.

NOTE *Despite the version numbers for future production code releases, Angular 2 and beyond is officially known as Angular.*

The Reason for Skipping Angular 3

The core Angular libraries are in `github.com/angular/angular`. All of them are versioned the same way, but distributed as different Node-based packages:

```
@angular/core            v2.3.0
@angular/compiler        v2.3.0
@angular/compiler-cli    v2.3.0
@angular/http            v2.3.0
@angular/router          v3.3.0
```

As you can see, the version number for the router package differs from the version number of the other packages. The simplest way to make the version numbers identical was to use version 4 instead of version 3, which in turn resulted in skipping Angular 3.

TypeScript Version

Upgrading the TypeScript dependency from v1.8 to v2.1 (or higher) does involve a breaking change. Fortunately, upgrading from Angular version 2 to version 4 (and beyond) does not involve a major rewrite, just a change in a few core libraries.

The Angular team has developed an automatic upgrade process, and the tool that they use might also become available for general use (possibly during 2017).

NOTE *The Angular 2 applications in this book use version 2.2.2 of the TypeScript compiler.*

Moving from Angular 2 to Angular

This section is optional if you are working with the latest release of Angular, which automatically includes all the features that were introduced in Angular 2.x beyond Angular 2.0.

A brief synopsis of several Angular 2.x releases is provided below because the new features were added to Angular 2 (but they are not part of Angular 2.0). In addition, the version of TypeScript has changed.

First, Angular 2.2 was released in November 2016; this version provides ahead-of-time (AOT) compilation compatibility and is discussed in more detail in Chapter 10. Next, Google released Angular 2.3 in December 2016, which includes the Angular Language Service for integrating with integrated development environments (IDEs). This service provides type completion and error-checking with Angular Templates. Moreover, object inheritance for components is featured in this service.

Regarding the version of TypeScript: Angular 2 used TypeScript 1.8, whereas the initial production release of Angular 4 uses TypeScript 2.1.6 (or higher), which involves some relatively small breaking changes.

What You Need to Learn for Angular Applications

Angular applications in this book are written in TypeScript, but you also need to acquire a basic proficiency in the following technologies:

- NodeJS (npm)
- ECMA5, ES6, and TypeScript
- Webpack 2.2 (or higher)

The following subsections contain more information about the items in the preceding list.

NodeJS

The code samples in this book require `node` v6.x.x (or higher) and `npm` 4.x.x (or higher). Determine the version on your machine with the following commands in a terminal:

```
node -v
npm -v
```

If necessary, navigate to the NodeJS home page to download a more recent version of the node executable. If you have not worked with Node, read an online tutorial to understand the purpose of the following commands:

- `npm install -g webpack`
- `npm install webpack@latest -save`
- `npm start`

ECMA5, ES6, and TypeScript

You need to learn the basic concepts of ES6 and TypeScript, and their respective home pages contain plenty of information to help you get started. In particular, learn about arrow functions, classes, template strings, and module loaders. Other useful features include the spread and rest operators (you will encounter them in Chapter 7). As you will see in subsequent chapters, Angular applications rely heavily on dynamic templates, which frequently involve interpolation (via the "{{}}" syntax) of variables.

Knowledge of ES6 is helpful if you plan to write Angular applications with TypeScript. Fortunately, the following website provides an online "playground" and links for documentation and code samples for TypeScript:

https://www.typescriptlang.org/play/

Familiarity with ECMA5 is also useful: for example, the `filter()` function is handy (e.g., with Angular Pipes), and the `map()` function can be useful when you combine `Observables` with HTTP requests in Angular applications. Other functions, such as `merge()` and `flatten()`, can also be useful, and you can learn about them and other functions on an as-needed basis.

You also need a basic understanding of Promises and Observables. Angular with TypeScript favors Observables, as do the code samples in this book, but you will encounter online code samples that use Promises. Avail yourself of online resources regarding `ECMA5`, Promises, and `Observables`.

Angular takes advantage of ES6 features such as components and classes, as well as features that are part of TypeScript, such as annotations and its type system. TypeScript is preferred over `ECMA5`or ES6 because (1) TypeScript supports all the features of ES6, and (2) TypeScript provides an optional type inferencing system that can catch many errors for you.

As for scaffolding tools, Webpack has become the de facto standard for Angular applications. Webpack (version 2.2 or higher) has become the de facto standard scaffolding tool for Angular applications, and its home page is located here:

https://webpack.github.io/

If you have not worked with Webpack, learn about the features of version 2.x (and bypass version 1.x).

Some additional relevant details: You can develop Angular applications in Electron, Webstorm, and Visual Studio Code. Check their respective websites for pricing and feature support. Finally, the following link includes style-related guidelines as well as "best practices" for developing Angular applications:

https://github.com/mgechev/angular2-style-guide

Glance through the preceding link to familiarize yourself with its content, and then you will know when to read the relevant sections as you progress through the chapters in this book.

A High-Level View of Angular

Angular was designed as a platform that supports Angular applications in a browser and supports server-side rendering and Angular applications on mobile devices. The first aspect—rendering Angular applications in browsers—is the focus of the chapters in this book. The second aspect—Angular Universal (aka server-side rendering)—is discussed briefly in Chapter 6. In essence, server-side rendering creates the "first view" of an Angular application on a server instead of a browser. Because browsers do not need to construct this view, they can render a view more quickly and create a faster perceived load time. The third aspect—Angular applications on mobile devices—is discussed in Chapter 8.

Angular also has a component-based architecture, where components are organized in a treelike structure (the same is true of Angular modules, as you will see in Chapter 5). Angular also supports powerful technologies that you will learn to become proficient in writing Angular applications. The simplest way to create an Angular application is to use the Angular CLI (discussed in detail later), which generates the required files for an Angular application from the command line.

Some of the important features of Angular are listed here:

- One-way data binding
- "Tree shaking"
- Change detection
- Style encapsulation

The first two features are briefly discussed below and the third feature is discussed in Chapter 2, where the code sample will make more sense to you than in this chapter.

Because the production version of Angular 4 was in March 2017, and additional "point" releases have been released (e.g., 4.1), it's possible that the latter releases will cause breaking changes in some of the code samples in this book. Consequently, you might need to modify application programming interface (API) calls in the affected code samples or change some `import` statements. You can always consult the online documentation regarding changes between consecutive releases of Angular:

https://angular.io/

One-Way Data Binding in Angular

Angular provides declarative one-way binding as the default behavior (but you can switch to two-way binding if you wish to do so). One-way binding acts as a unidirectional change propagation that provides two advantages over two-way data binding: (1) an improvement in performance because of eliminating the `$digest` cycle and watchers in Angular 1.x, and (2) a reduction in code complexity. Angular also supports stateful, reactive, and immutable models. The meaning of the previous statement will become clearer as you work with Angular applications.

Angular applications involve defining a top-level ("root") module that references a `Component` that in turn specifies an HTML element (via a mandatory `selector` property) that is the "parent" element of the `Component`. The definition of the `Component` involves a so-called "decorator," which includes a `selector` property and a `template` property (or a `templateUrl` property).

The `template` property includes a mixture of HTML and custom markup (which can be placed in a separate file and then referenced via the `templateUrl` property). In addition, the `Component` is immediately followed by a TypeScript class definition that includes "backing code" that is associated with component-related variables that appear in the `template` property. These details will become much clearer after you have worked with some Angular applications.

NOTE *The `templateUrl` property and `styleUrls` property refer to files, whereas the `template` property and `styles` property refer to inline code.*

Tree Shaking an Angular Application

Tree shaking refers to "pruning" the unnecessary files from an Angular project (other technologies have the same concept), just like shaking the dead branches from a tree, in order to create a production version of an Angular application. This production version can be significantly smaller, and hence require less loading time. Webpack 2 and Angular AOT support tree shaking, and you will see an example with AOT in Chapter 10.

A High-Level View of Angular Applications

Angular applications consist of a combination of built-in components and custom components (the latter are written by you), each of which is typically defined in a separate TypeScript file (with a `ts` extension).

Each component specifies its dependencies via `import` statements. There are various types of dependencies available in Angular, such as directives and pipes (discussed later in this chapter).

A *custom directive* is essentially the contents of a TypeScript file that defines a component. Thus, a custom directive consists of `import` statements, a `Component` decorator, and an exported TypeScript class.

Angular provides *built-in directives*, such as `*ngIf` (for "if" logic) and `*ngFor` (for loops). These two directives are also called *structural directives* because they modify the content of an HTML page.

Angular *built-in pipes* include date and numeric items (currency, decimal, number, and percent), whereas *custom pipes* are defined by you.

In addition, TypeScript classes use a *decorator* (which is a built-in function) that provides metadata to a class, its members, or its method arguments. Decorators are easy to identify because they always have the `@` prefix. Angular provides a number of built-in decorators, such as `@Component()` and `@NgModule`.

This concludes the high-level introduction to Angular features. The next portion of this chapter introduces the Angular CLI, which is used throughout this book to create Angular applications.

The Angular CLI

During the beta releases of Angular, developers used a manual process to create applications, or they used a "starter" or "seed" project (often available on GitHub). However, those projects are often out of date after new releases of Angular are available. Hence, the code samples in this book are based on the Angular CLI, which is the official Angular application generator from Google.

The Angular CLI is a command-line tool called `ng` that generates complete Angular applications, including test-related code, and (by default) also launches `npm install` to install required files in `node_modules`. The home page for the Angular CLI is located here:

https://cli.angular.io

The Angular CLI generates a configuration file called `package.json` to manage the necessary dependencies and their version numbers. After generating an Angular application, navigate to the `node_modules` subdirectory, and you will see an assortment of Angular subdirectories that contain files that are required for Angular applications.

The Angular CLI is a superior alternative to using "starter" projects or creating projects manually, and its feature set will continue to improve over time.

Installing the Angular CLI

You need to perform several steps to install the Angular CLI. First uninstall the previous CLI (if you installed an older version) with this command:

```
sudo npm uninstall -g angular-cli
npm cache clean
```

Next, install the new CLI with this command (note the new package name):

```
[sudo] npm install -g @angular/cli
```

Create a new project called `hello-world` with Angular 4 as follows:

```
ng new hello-world
```

The Angular CLI provides everything except your custom code, and also requires noticeably more time to install than starter applications. Second, the Angular CLI enables you to generate new components, routers, and so forth, which are possible with starter applications. Third, the Angular CLI is based purely on TypeScript, and the generated application includes the JavaScript Object Notation (JSON) files `tsconfig.json`, `tslint.json`, `typedoc.json`, and `typings.json`. On the other hand, the starter applications tend to use Webpack, which involves the configuration file `webpack.config.js`, which includes information for Angular applications.

Features of the Angular CLI

The `ng` executable supports various options, and some command-line invocations are shown here:

```
ng new app-root-name
ng build
ng deploy
```

```
ng e2e
ng generate <component-type>
ng generate route
ng generate ...
ng lint
ng serve
ng test
ng x18n
```

The `ng g` option is equivalent to the `ng generate` option, which enables you to generate an Angular custom `Component`, an Angular `Pipe` (discussed in Chapter 5), and so forth. The `ng x18n` option extracts i18n messages from source code. The next section shows you an example of generating an Angular custom `Component` in an application, and the contents of the files that are automatically generated for you.

The default prefix is `app` for components (e.g., `<app-root></app-root>`), but you can specify a different prefix with this invocation:

```
ng new app-root-name -prefix abc
```

NOTE *Angular applications created via* `ng` *always contain the* `src/app` *directory.*

Information about upgrading the Angular CLI is located here:

https://github.com/angular/angular-cli

Documentation for the Angular CLI is located here:

http://cli.angular.io

Now that you have an understanding of some of the features of the `ng` utility, let's create our first Angular application, which is the topic of the next section.

A "Hello World" Application via the Angular CLI

Navigate to a convenient directory and create an Angular application called `myapp` with the following command:

```
ng new myapp
```

NOTE *Earlier versions of ng require an* `--ng4` *switch.*

After the preceding command has completed, navigate inside the new project:

```
cd myapp
```

Launch the Angular application with the following command:

```
ng serve
```

The preceding command automatically launches a browser session at the following URL:

```
localhost:4200
```

You will see the following displayed in your screen:

app works!

The preceding string is specified in the `template` property in the file `app/app.component.ts.`

NOTE *The `template` property (or the `templateUrl` property) is where you will place custom code that will generate `HTML` output, which is then inserted into the `<app-root>` element in the `index.html` Web page.*

The Structure of an Angular Application

When you invoke the `ng` command to create an Angular application, here are the files that are automatically created:

```
angular-cli.json
dist/ (details omitted)
e2e/  (details omitted)
karma.conf.js
node_modules
package.json
protractor.conf.js
README.md
src/app/app.component.css
src/app/app.component.html
src/app/app.component.spec.ts
src/app/app.component.ts
src/app/app.module.ts
src/environments/environment.prod.ts
src/environments/environment.ts
src/favicon.ico
src/index.html
src/main.ts
src/polyfills.ts
src/styles.css
src/test.ts
src/tsconfig.json
tslint.json
```

The `dist` subdirectory includes the compiled project and JavaScript bundle, whereas the `e2e` subdirectory includes files for end-to-end testing (invoke the command `ng test` to execute the tests).

The file `app.component.ts` (shown in bold) includes custom TypeScript code in the code samples in this book, and (to a lesser extent) you might need to update the file `app.module.ts`. You will modify `index.html` when you need to include JavaScript `<script>` elements (for jQuery, Bootstrap, and so forth), and the `styles.css` file is for global Cascading Style Sheets (CSS) style rules (if any).

Finally, the files `angular-cli.json`, `package.json`, and `tslint.json` are configuration files for `ng`, `npm`, and `tslint`, respectively. Note that when you invoke the command `ng lint`, it invokes the `tslint` utility, which in turn relies on the file `tslint.json`.

The Naming Convention for Angular Project Files

The files in the `app` subdirectory have a naming convention that comprises three parts: the type of functionality, whether it's a component or module, and a suffix that indicates the type of code. For example, the TypeScript file `app.component.ts` includes component-related code for the application, whereas the TypeScript file `app.module.ts` includes module-related code. In addition, the file `app.component.css` contains CSS selectors for the component, and `app.component.html` contains HTML markup for the same component.

The next portion of this chapter contains two sections: The first part discusses the contents of `index.html` and various JavaScript configuration related files, and the second part discusses application-related files that are contained in the `app` subdirectory. Alas, reading these sections is a "long slog," and although it's recommended that you read them, feel free to skim this section and return after you have launched your first Angular application. There are trade-offs with both reading styles, so proceed with this material in the manner that best suits your learning style.

Now let's take a look at the contents of the HTML Web page `index.html`, which is the main Web page for our Angular application.

The index.html Web Page

Listing 1.1 displays the contents of `index.html` for a new Angular application that is generated from the command line via the `ng` utility.

LISTING 1.1: index.html

```
<!doctype html>
<html>
<head>
  <meta charset="utf-8">
  <title>Angular Application</title>
  <base href="/">

  <meta name="viewport" content="width=device-width,
                                        initial-scale=1">
  <link rel="icon" type="image/x-icon" href="favicon.ico">
</head>
<body>
  <app-root>Loading...</app-root>
</body>
</html>
```

Listing 1.1 is minimalistic: Only the custom `<app-root>` element (which you will see in the `selector` property in `app/app.component.ts`) gives you an indication that this Web page is part of an Angular application.

NOTE *The JavaScript dependencies are dynamically inserted in* `index.html` *by the Angular CLI during the "build" of the project.*

Before we delve into the TypeScript files in an Angular application, let's take a quick detour to understand how `import` statements work in Angular applications. Feel free to skip the next section if you are already familiar with `import` and `export` statements in Angular.

Exporting and Importing Packages and Classes (Optional)

Keep in mind the following point: Every TypeScript class that is imported in a TypeScript file must be exported in the TypeScript file where that class is defined. You will see many examples of `import` and `export` statements; in fact, this is true of every Angular application in this book.

There are 2 common types of `import` statements: one type involves importing packages from Angular modules, and the other type involves importing custom classes (written by you). Here is the syntax for both types:

```
import {some-package-name} from 'some-angular-module';
import {some-class }       from 'my-custom-class';
```

Here is an example of both types of `import` statements:

```
import { NgModule }          from '@angular/core';
import {EmpComponent}        from './emp.component';
```

In the preceding code snippet, the `NgModule` package is imported from the `@angular/core` module that is located in the `node_modules` directory. The `EmpComponent` class is a custom class that is defined and exported in the TypeScript file `emp.component.ts`.

In the second `import` statement, the "`./`" prefix is required when a custom class is imported from a TypeScript file, and notice the omission of the "`.ts`" suffix.

The next several sections discuss three application-related TypeScript files in the `src/app` subdirectory: `main.ts`, `app.component.ts`, and `app.module.ts`. These files are the bootstrap file, the main module, and the main component class, respectively, for this Angular application.

Here is the condensed explanation of the purpose of these three files: Angular uses `main.ts` as the initial "entry point" to bootstrap the Angular module `AppModule` (defined in `app.module.ts`), which in turn includes the component `AppComponent` (defined in `app.component.ts`), as well as any other custom components (and modules) that you have imported into `AppModule`.

Moreover, these three files are located in the `src/app` subdirectory, which is also where you place custom components (and modules), or some suitable subdirectory of `src/app` whose name is based on its feature.

The Bootstrap File `main.ts`

Listing 1.2 displays the contents of `main.ts` in the `src` subdirectory (not the `src/app` subdirectory) that imports and bootstraps the top-level Angular module `AppModule`.

LISTING 1.2: main.ts

```
import { enableProdMode } from '@angular/core';
import { platformBrowserDynamic } from '@angular/platform-
                                             browser-dynamic';

import { AppModule } from './app/app.module';
import { environment } from './environments/environment';
```

```
if (environment.production) {
  enableProdMode();
}

platformBrowserDynamic().bootstrapModule(AppModule);
```

The first line of code in Listing 1.2 is an `import` statement that is needed for the conditional logic later in the code listing. The second `import` statement that you will see in many Angular code samples, and it's necessary for launching Angular applications on desktops and laptops.

The third `import` statement involves the top-level module of Angular applications, which in turn contains all the custom components and services that are included in this Angular module. The fourth `import` statement contains environment-related information that is used in the next conditional logic snippet: if the current application is in production mode, the `enableProdMode()` function is executed.

The final line of code is the actual bootstrapping process, which involves rendering the code in `app.component.ts` in a browser.

The Top-Level Module File `app.module.ts`

Listing 1.3 displays the contents of `app.module.ts` (located in the `src/app` subdirectory) that exports the top-level Angular module `AppModule`.

LISTING 1.3: app.module.ts

```
import { BrowserModule } from '@angular/platform-browser';
import { NgModule }      from '@angular/core';
import { FormsModule }   from '@angular/forms';
import { HttpModule }    from '@angular/http';
import { AppComponent }  from './app.component';

@NgModule({
  declarations: [
    AppComponent
  ],
  imports: [
    BrowserModule,
    FormsModule,
    HttpModule
  ],
  providers: [],
  bootstrap: [AppComponent]
})
export class AppModule { }
```

Listing 1.3 includes two `import` statements that import `BrowserModule`, `NgModule`, `FormsModule`, `HttpModule`, and `BrowserModule`, all of which are part of Angular. The last `import` statement imports the class `AppComponent`, which is the top-level component illustrated in Listing 1.4.

NOTE *Angular dependencies always contain the @ symbol, whereas custom dependencies specify a relative path to TypeScript files.*

Next, the `@NgModule` decorator includes an object with various properties (discussed in the next section). These properties specify metadata for the class `AppModule`, which is exported in the final line of code. The metadata in `AppModule` involves the following properties, each of which is an array of values: `imports`, `providers`, `declarations`, `exports`, and `bootstrap`.

In Listing 1.3, the array properties `declarations`, `imports`, and `bootstrap` are non-null, whereas the `providers` property is an empty array. This metadata is required for Angular to bootstrap the code in `AppComponent`, which in turn contains the details of what is rendered (e.g., an `<h1>` element) and where it is rendered (e.g., the `app-root` element in `index.html`).

Now let's take a closer look at the purpose of each array-based property in the `@NgModule` decorator to understand their purpose.

The MetaData in `@NgModule`

The `imports` array includes *modules* that are required for this application, such as `BrowserModule`, and some optional modules (e.g., `FormModule`) for this application. The `imports` array is not transitive: if module A imports module B and module B imports module C, then module C is not imported into module A.

Next, the `providers` array is an array of application-wide services for this Angular application. A service is something that provides behind-the-scenes functionality that is not visible in the application. The `providers` array includes any injectable services that have been defined via the `@Injectable` decorator (discussed later).

The `declarations` array consists of *components* that are required for this application, such as the `AppComponent`, and also custom components, directives, and pipes, all of which have (module-level) private scope. Specifically, *private scope* means that everything in the `declarations` array is accessible only via other components, directives, and pipes that are declared in the same module.

Keep in mind that components listed in the `declarations` property *must* be exported from their respective component-related files. For example, `AppComponent` is exported from `app.component.ts` (shown later), which enables you to import it in `app.module.ts` and also specify it in the `declarations` property.

The `exports` array includes components, directives, and pipes that are required in other components in the application.

The `schemas` array includes the value `CUSTOM_ELEMENTS_SCHEMA`, which provides additional information that is "external" to the application; see the example in Chapter 2 and the example in Chapter 5.

Finally, the `bootstrap` array includes the name of the component that will be "bootstrapped" when the Angular application is launched in a browser.

NOTE *In general, it's advisable to bootstrap only one component via the* `bootstrap` *property.*

The Top-Level Component File app.component.ts

Listing 1.4 displays the contents of `app.component.ts` (located in the `src/app` subdirectory) that exports the top-level Angular component `AppComponent`.

LISTING 1.4: app.component.ts

```
import { Component } from '@angular/core';

@Component({
  selector:    'app-root',
  templateUrl: './app.component.html',
  styleUrls:   ['./app.component.css']
})
export class AppComponent {
  title = 'app works!';
}
```

Listing 1.4 starts with an `import` statement for the Angular `@Component` decorator to define metadata for the class `AppComponent`. At a minimum, the metadata involves two properties: `selector` and `template`.

Except for routing-related components, both of these properties are required in custom components. In this example, the `selector` property specifies the custom element `app-root` (which you can change) that you saw in the HTML Web page `index.html`.

The `template` property specifies the HTML markup that will be inserted in the custom element `app-root`. In this example, the markup is an `<h1>` element containing some text. The final line of code in Listing 1.4 is an `export` statement that makes the `AppComponent` class available for import in other TypeScript files, such as `app.module.ts`, which is shown in Listing 1.3.

NOTE *The files `main.ts`, `app.component.ts`, and `app.module.ts` are in the `src/app` subdirectory for all the Angular projects in this book.*

Launch your Angular application by navigating to the root directory and entering the following command from a shell:

```
ng serve
```

The next section discusses Angular template syntax, which you will use in your custom code in the `template` property.

A Simple Angular Template

As you saw in Listing 1.4, the file `app.component.ts` includes a `template` property with an `<h1>` element that contains a single line of text. However, Angular enables you to specify multiple lines of text when you place everything inside a pair of matching backticks ("`"). This syntax (introduced in ES6) is used heavily in Angular applications, and it conveniently supports variable interpolation.

Listing 1.5 displays the contents of a new `app.component.ts` file, which has more content than Listing 1.4. This code sample illustrates how to use interpolation in a `template` property.

LISTING 1.5: app.component.ts

```
import { Component } from '@angular/core';

@Component({
  selector: 'app-root',
  template: `
              <h3>Hello everyone!</h3>
```

```
            <h3>My name is {{emp.fname}} {{emp.lname}}</h3>
          `
})
export class AppComponent {
  public emp = {fname:'John',lname:'Smith',city:'San Francisco'};
}
```

Listing 1.5 includes two `<h3>` elements; the first is a simple text string and the second includes "curly braces" to reference data values that are defined in the class `AppComponent`. Angular uses something called *interpolation* with the elements of the literal object `emp` and then substitutes the variables inside the curly braces with their actual values in the second `<h3>` element.

Working with Components in Angular

An Angular application is a tree of nested components, where the top-level component is the application. The components define the user interface (UI) elements, screens, and routes. In general, organize Angular applications by placing each custom component in a TypeScript file and then import that same TypeScript file in the "main" file (which is often named `app.component.ts`), which includes the top-level component.

The Metadata in Components

Angular components are often a combination of an `@Component` decorator and a class definition that can optionally contain a constructor. A simple example is shown here:

```
import { Component }  from '@angular/core';
import {EmpComponent} from './emp/emp.component';

@Component({
    selector:   'app-container',
    template:   `{{message}}<tasks></tasks>`,
    directives: [EmpComponent]
})
```

The preceding `@Component` decorator includes several properties, some of which are mandatory and others are optional. Let's look at both types in the preceding code block.

The `selector` property is mandatory, and it specifies the HTML element (whether it's an existing element or a custom element) that serves as the "root" of an Angular application.

Next, the `template` property (or a `templateUrl` property) is mandatory, and it includes a mixture of markup, interpolated variables, and TypeScript code. One important detail: the `template` property requires backticks when its definition spans multiple lines.

The `directives` property is an optional property that specifies an array of components that are treated as nested components. In this example, the `directives` property specifies the component `EmpComponent`, which is also imported (via an `import` statement) near the beginning of the code block. Notice that the `import` statement does not contain the @ symbol, which means that `EmpComponent` is a custom component defined in the file `emp/emp.component.ts`.

Stateful versus Stateless Components in Angular

In high-level terms, a *stateful* component retains information that is relevant to other parts of the same Angular application. On the other hand, stateless components do not maintain an application state, nor do they request or fetch data: they are passed data via property bindings from another component (such as its parent).

The code samples in this book are usually a combination of stateful components, stateless components, and sometimes "value objects," which are instances of custom classes that model different entities (such as an employee, customer, student, and so forth).

You will see an example of a presentational component in Chapter 2. In the meantime, a good article that delves into stateful and stateless components is located here:

https://toddmotto.com/stateful-stateless-components#stateful

Generating Components with the Angular CLI

Earlier you saw some of the options for the `ng` utility, and this section discusses how to use the `generate` option. In addition, here are some self-explanatory examples:

```
ng generate component  mycomp
ng generate directive  mydir
ng generate pipe       mypipe
ng generate route      myroute
ng generate service    mycomp
```

The following commands are equivalent to their counterparts in the preceding list, except that they use short aliases:

```
ng g c mycomp
ng g d mydir
ng g p mypipe
ng g r myroute
ng g s mycomp
```

The preceding commands create new files that are located in new subdirectories of `src/app`. For example, the command `ng g c mycomp` creates the subdirectory `src/app/mycomp` and populates it with files (discussed later).

If you do not want to create a new subdirectory, use the option `-flat` and the new files will be placed in `src/app` instead of `src/app/mycomp`.

Now let's invoke the following command in an Angular application to generate a `student` component:

```
ng g c student
```

After the preceding command has completed, you will see the following output:

```
installing component
  create src/app/student/student.component.css
  create src/app/student/student.component.html
  create src/app/student/student.component.spec.ts
  create src/app/student/student.component.ts
```

As you can see, the preceding output displays four new files: a CSS file, an HTML file, a TypeScript test file, and a component definition file.

Listing 1.6 displays the contents of `student.component.ts` that contains the code for the `student` component.

LISTING 1.6: student.component.ts

```
import { Component, OnInit } from '@angular/core';

@Component({
  selector: 'app-student',
  templateUrl: './student.component.html',
  styleUrls: ['./student.component.css']
})
```

```
export class StudentComponent implements OnInit {

  constructor() { }

  ngOnInit() {
  }
}
```

NOTE *When you use `ng` to create a custom component, `ng` inserts an `import` statement in `app.module.ts` and updates the `declarations` property with a reference to the new custom component.*

For your convenience, `ng` also generates a test file when you generate a component from the command line. In particular, Listing 1.7 displays the contents of the test file called `student.component.spec.ts`.

LISTING 1.7: student.component.spec.ts

```
/* tslint:disable:no-unused-variable */

import { TestBed, async } from '@angular/core/testing';
import { StudentComponent } from './student.component';

describe('Component: Student', () => {
  it('should create an instance', () => {
    let component = new StudentComponent();
    expect(component).toBeTruthy();
  });
});
```

Listing 1.7 includes `import` statements to make various components available to this test file, including the `StudentComponent` that is defined in Listing 1.6.

The autogenerated contents of `student.component.html` are shown here:

```
<p>
  student works!
</p>
```

As you can see, the file `student.component.html` is minimalistic, and you are free to add application-related HTML markup in this file.

Note that the file `student.component.css` is currently empty; add any CSS selectors that you need for styling purposes.

One other detail: ng enables you to create a component inside an existing component. For example, the following command creates a profile component inside the student component:

```
ng g component student/profile
```

You must invoke the preceding command from the student subdirectory, otherwise you will see the following error message:

```
You have to be inside an angular-cli project in order to use
the generate command.
```

When you are ready, create a production build of an Angular application with the following command:

```
ng build --prod
```

Keep in mind that you can significantly reduce the size of an Angular application via AOT, which is discussed in more detail in Chapter 10.

Syntax, Attributes, and Properties in Angular

Angular introduced the square brackets "[]" notation for attributes and properties, as well as the parentheses "()" notation for functions that handle events. This new syntax is actually valid HTML5 syntax. Here is an example of a code snippet that specifies an attribute and a function:

```
<foo [bar]= "x+1" (baz)="doSomething()">Hello World</foo>
```

An example that specifies a property and a function is shown here:

```
<button [disabled]="!inputIsValid" (click)="authenticate()"
                                                  >Login </button>
```

An example of a data-related element with a custom element is shown here:

```
<my-chart [data]="myData" (drag)="handleDrag()"></my-chart>
```

The new syntax in the preceding code snippet eliminates the need for many built-in directives, as you will see later in this chapter.

Attributes versus Properties in Angular

Keep in mind the following distinction: A property can specify a complex model, whereas an attribute can only specify a string. For example, in Angular 1.x you can write the following:

```
<my-directive foo="{{something}}"></my-directive>
```

The corresponding code in Angular (which does not require interpolation) is shown here:

```
<my-directive [foo]="something"></my-directive>
```

The new architecture for Angular provides improved performance and a mechanism for developing "cleaner" Angular applications that can be developed, enhanced, and maintained more quickly.

The next section contains a code sample involving a `<button>` element, which is probably one of the most common UI controls in HTML Web pages.

Displaying a Button in Angular

After having soldiered through all the code listings in this chapter, and also reading explanations about their purpose, you might be wondering if application development in Angular is going to be a long and tedious process. Fortunately, you can create many basic applications with a small amount of code. When you are ready to create medium-sized applications, you can take advantage of the component-based nature of Angular applications to incrementally add new components (and modules).

As a simple example, the file `app.component.ts` in this section includes all the custom code for this Angular application. The `ng` utility was used to generate all the files in this code sample, and you won't need to tinker manually with those other files.

Copy the directory `ButtonClick` from the companion disc into a convenient location. Listing 1.8 displays the contents of `app.component.ts` that illustrates how to render a `<button>` element and respond to click events by displaying the number of times that users have clicked the `<button>` element during the current session.

LISTING 1.8: app.component.ts

```
import { Component } from '@angular/core';

@Component({
   selector: 'app-root',
   template: `<div>
               <button (click)="clickMe()">ClickMe</button>
               <p>Click count is now {{clickCount}}</p>
              </div>`,
   styles: [` button {
                 color: red;
              }`
           ]
})
export class AppComponent {
   clickCount = 0;

   clickMe() {
      ++this.clickCount;
      console.log("click count: "+this.clickCount);
   }
}
```

Listing 1.8 starts with an `import` statement that appears in `app.component.ts`, which you will see in every code sample in this book. This statement gives you access to the `Component` decorator, which injects metadata into the TypeScript class called `AppComponent`. When the TypeScript compiler transpiles (converts) the TypeScript code into ECMA5, the metadata will also be included to run the Angular application in modern browsers.

The required `selector` property specifies a value of `app-root`, which is the custom element (listed in `index.html`) that serves as the "container" where the content of the `template` element is rendered.

In this example, the `template` property includes a `<button>` element that responds to click events and a `<p>` element whose contents are updated when users click the `<button>` element. As you can see, the value of the term `(click)` is the `clickMe()` function (defined in the `AppComponent` class), which increments and then displays the value of the `clickCount` variable.

In addition, the `styles` property specifies a value of `red` for the `<button>` element. The `styles` property is an example of component style, which means that the styles only apply to the template of the given component.

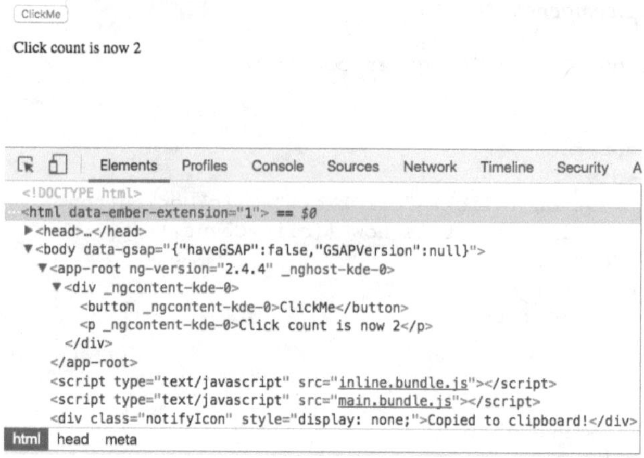

ClickMe

Click count is now 2

```
Elements  Profiles  Console  Sources  Network  Timeline  Security  A
<!DOCTYPE html>
<html data-ember-extension="1"> == $0
 ▶<head>…</head>
 ▼<body data-gsap="{"haveGSAP":false,"GSAPVersion":null}">
   ▼<app-root ng-version="2.4.4" _nghost-kde-0>
     ▼<div _ngcontent-kde-0>
         <button _ngcontent-kde-0>ClickMe</button>
         <p _ngcontent-kde-0>Click count is now 2</p>
       </div>
     </app-root>
     <script type="text/javascript" src="inline.bundle.js"></script>
     <script type="text/javascript" src="main.bundle.js"></script>
     <div class="notifyIcon" style="display: none;">Copied to clipboard!</div>
html  head  meta
```

FIGURE 1.1 A <button> Element that Responds to Click Events.

In effect, Angular applies CSS locally instead of globally by generating unique attributes that are visible when you click the `Elements` tab in Chrome Web Inspector.

More detailed information regarding component styles in Angular is located here:

https://angular.io/docs/ts/latest/guide/component-styles.html

The next portion of Listing 1.8 is the definition of the `AppComponent` class that includes the `clickCount` variable, which is incremented in the `clickMe()` function.

Navigate to the `src` subdirectory of this application and invoke the following command:

```
ng serve
```

Figure 1.1 displays the browser output from this Angular application. You can also see the file `main.bundle.js`, which includes minified "tree shaken" code, which is discussed in more detail in Chapter 10.

Element versus Property

In Listing 1.8, the `selector` property matched the element <app-root></app-root> in the HTML page `index.html`:

```
selector: 'app-root'
```

However, you can also specify a property instead of an element. For example, suppose that `index.html` includes the following element:

```
<div app-root>Loading. . .</div>
```

You also need to modify the `selector` property as follows:

```
selector: '[app-root]'
```

Summary

This chapter started with a description of the Angular version numbers, prerequisites for Angular, and an overview of Angular and its hierarchical component-based structure.

Next you learned about the Angular CLI utility `ng` and how to create an Angular "Hello world" application with the `ng` utility. You also learned about the TypeScript files `main.ts`, `app.component.ts`, and `app.module.ts`, which contain TypeScript code for an Angular application. Next you learned about the reason for transpiling the code in an Angular application into ECMA5. Finally, you saw the code for an Angular application that displays a `<button>` element that also responds to click events.

CHAPTER 2

UI CONTROLS AND USER INPUT

This chapter contains Angular applications with an assortment of user interface (UI) Controls and examples for handling user interaction, such as user input and mouse events. Keep in mind that the code samples in this chapter render UI Controls using standard HTML syntax instead of using functionality that is specific to Angular. In addition, the last section in this chapter contains links to toolkits that provide Angular UI components.

The first part of this chapter briefly discusses basic debugging techniques that you can use in any Angular application. This section uses the button-related code sample in Chapter 1 to illustrate how to use the debugger in Angular.

The second part of this chapter shows you how to manage lists of items, which includes displaying, adding, and deleting items from a list. Note that forms in Angular are deferred until Chapter 4, where you will also learn about `Controls` and `ControlGroups`.

The third section of this chapter contains two examples of displaying a list of user names: the first retrieves user names that are stored as strings in a JavaScript array, and the second retrieves user names that are stored in object literals in a JavaScript array. The third section goes a step further: You will learn how to define a custom user component that contains user-related information (also contained in a JavaScript array).

NOTE *When you copy a project directory from the companion disc, if the node_modules directory is not present, then copy the top-level node_modules*

directory that has been soft-linked inside that project directory (which is true for most of the sample applications).

Now let's learn how to perform some basic debugging in Angular.

Debugging Angular Code in the Console

This section shows you how to use `ng.probe()` to "step through" the execution of an Angular application, and to find (or update) the values of variables. The information in this section will help you detect simple errors in your applications. In case you're interested, a much more powerful debugger is Augury (a Chrome extension), which is discussed in Chapter 10.

Now launch the Angular application `ButtonClick` in Chapter 1 (Listing 1.8) in Chrome and perform the following steps:

1) Open Chrome Web Inspector.
2) Click the `Elements` tab.
3) Click the `<app-root>` element.

Now click the `Console` tab and enter the following snippet in the console:

```
ng.probe($0)
```

You will see a `DebugElement` that looks something like the following:

```
DebugElement {nativeNode: app-root, parent: null, listeners:
Array[0], providerTokens: Array[1], properties: Map…}
```

Expand the preceding object and peruse its elements. Click the button (at the top of the screen) several times, and then obtain a reference to the instance of the component class with the following code snippet:

```
ng.probe($0).componentInstance
```

The preceding snippet displays the following element:

```
UserInput {clickCount: 3}
```

Modify the value of `clickCount` with the following code snippet:

```
ng.probe($0).componentInstance.clickCount = 7
```

The preceding code snippet changes the value of `clickCount` to `7`, which admittedly does not have any practical purpose in this code sample. However, in other Angular applications that contain various input fields and widgets, this functionality could be useful for testing purposes.

The following code snippet provides component-related information:

```
ng.probe($0).injector._depProvider.componentView
```

The preceding snippet displays the following output:

```
AppView {proto: AppProtoView, renderer: DebugDomRenderer,
        viewManager: AppViewManager_, projectableNodes: null,
                             containerAppElement: AppElement…}
ng.probe($0).injector._depProvider.componentView.
                                                changeDetector
```

The next code snippet provides additional information:

```
ChangeDetector_UserInput_0 {id: "UserInput_0"",
numberOfPropertyProtoRecords: 2, bindingTargets: Array[1],
directiveIndices: Array[0], strategy: 5…}
```

Experiment with the available elements by expanding them and inspecting their contents. In addition, the following link contains useful information about Chrome development tools:

https://developers.google.com/web/tools/chrome-devtools/javascript

Now let's create a simple Angular application that shows you how to display a list of strings via the `ngFor` directive, as discussed in the next section.

The `ngFor` Directive in Angular

The code sample in this section displays a hard-coded list of strings via the `*ngFor` directive. This very simple code sample is a starting point from which you can create more complex—and more interesting—Angular applications.

Copy the directory `SimpleList` from the companion disc into a convenient location. Listing 2.1 displays the contents of `app.component.ts` that illustrates how to display a list of items using the `*ngFor` directive in Angular.

LISTING 2.1: app.component.ts

```
import {Component} from '@angular/core';

@Component({
  selector: 'app-root',
  template: `<div *ngFor="let item of items">
               {{item}}
             </div>`
})
export class AppComponent {
  items = [];

  constructor() {
    this.items = ['one','two','three','four'];
  }
}
```

Listing 2.1 contains a `Component` annotation that in turn contains the standard `selector` property. The `template` property consists of a `<div>` element. This element contains the `ngFor` directive, which iterates through the `items` array and displays each item in that array. Notice that the `items` array is initialized as an empty array in the `AppComponent` class, and then its value is set to an array of four strings in the `constructor` method.

Launch the application in this section and you will see the following output in a browser session:

one
two
three
four

Now that you understand how to display items in an array, let's take a short digression to learn about the type of Angular code that can keep track of the radio button that users have clicked ("checked"). After that we'll see how to use a `<button>` element to add new user names to a list of users.

Angular and Radio Buttons

Copy the directory `RadioButtons` from the companion disc into a convenient location. Listing 2.2 displays the contents of `app.component.ts` that illustrates how to render a set of radio buttons and keep track of which button users have clicked.

LISTING 2.2: app.component.ts

```
import {Component} from '@angular/core';

@Component({
  selector: 'app-root',
  template: `
   <h2>{{radioTitle}}</h2>
   <label *ngFor="let item of radioItems">
      <input type="radio" name="options"
             (click)="model.options = item"
             [checked]="item === model.options">
      {{item}}
   </label>
   <p><button (click)="model.options='option1'">Set Option
                                                        #1</button>
  `
})
export class AppComponent {
  radioTitle = "Radio Buttons in Angular";
  radioItems = ['option1','option2','option3','option4'];
  model = { options: 'option3' };
}
```

Listing 2.2 defines the `AppComponent` component whose `template` property contains three parts: a `<label>` element, an `<input>` element, and a `<button>` element. The `<label>` element contains an `ngFor` directive that displays a set of radio buttons by iterating through the `radioItems` array that is defined in the `AppComponent` class.

By default, the first radio button is highlighted. However, when users click the `<button>` element, the `(click)` attribute of the `<input>` element sets the *current* item to the value of `model.options`, and then the `[checked]` attribute of the `<input>` element sets the *checked* item to the current value of `model.options`. As you can see, the `<input>` element in Listing 2.2 contains functionality that is more compact than using JavaScript to achieve the same results.

Adding Items to a List in Angular

Copy the directory `AddListButton` from the companion disc into a convenient location. Listing 2.3 displays the contents of `app.component.ts` that illustrates how to append strings to an array of items when users click a button.

LISTING 2.3: app.component.ts

```
import {Component} from '@angular/core';

@Component({
   selector: 'app-root',
   template: `
       <div>
         <input #fname>
         <button (click)="clickMe(fname.value)">ClickMe
                                                </button>
         <ul>
           <li *ngFor="let user of users">
             {{user}}
           </li>
         </ul>
       </div>`
})
export class AppComponent {
   users = ["Jane", "Dave", "Tom"];

   clickMe(user) {
      console.log("new user = "+user);
      this.users.push(user);
/*
      // prevent empty user or duplicates
      if(user is non-null) {
        if(user is duplicate) {
          // display alert message
        } else {
          // display alert message
        }
      } else {
        // display alert message
      }
*/
   }
}
```

Listing 2.3 contains code that is similar to Listing 2.1 (which displays a list of strings). In addition, the `template` property in Listing 2.3 contains an `<input>` element so that users can enter text. When users click the `<button>` element, the `clickMe()` method is invoked with `fname.value` as a parameter, which is a reference to the text in the `<input>` element.

Notice the use of the `#fname` syntax as an identifier for an element, which in this case is an `<input>` element. Thus, the text that users enter in the

<input> element is referenced via `fname.value`. The following code snippet provides this functionality:

```
<input #fname>
<button (click)="clickMe(fname.value)">ClickMe</button>
```

The `clickMe()` method in the `AppComponent` component contains a `console.log()` statement to display the user-entered text (which is optional) and then appends that text to the array `user`. The final section in Listing 2.3 consists of a commented out block of pseudocode that prevents users from entering an empty string or a duplicate string. This code block involves "pure" JavaScript, and the actual code is left as an exercise for you.

Deleting Items from a List in Angular

This section enhances the code in the previous section by adding a new <button> element next to each list item.

Copy the directory `DelListButton` from the companion disc into a convenient location. Listing 2.4 displays the contents of `app.component.ts` that illustrates how to delete individual elements from an array of items when users click a button that is adjacent to each array item.

LISTING 2.4: app.component.ts

```
import {Component} from '@angular/core';

@Component({
   selector: 'app-root',
   template: `
     <div>
       <input #fname>
       <button (click)="clickMe(fname.value)">ClickMe</button>
       <ul>
         <li *ngFor="let user of users">
           <button (click)="deleteMe(user)">Delete</button>
           {{user}}
         </li>
       </ul>
     </div>`
})
export class AppComponent {
```

```
    users = ["Jane", "Dave", "Tom"];

    deleteMe(user) {
       console.log("delete user = "+user);
       var index = this.users.indexOf(user);

       if(index >=0 ) {
          this.users.splice(index, 1);
       }
    }
    clickMe(user) {
       console.log("new user = "+user);
       this.users.push(user);
 /*
       // prevent empty user or duplicates
       if(user is non-null) {
         if(user is duplicate) {
           // display alert message
         } else {
           // display alert message
         }
       } else {
         // display alert message
       }
 */
    }
 }
```

Listing 2.4 contains an `ngFor` directive that displays a list of “pairs” of items, where each “pair” consists of a `<button>` element followed by a user in the `users` array.

When users click any `<button>` element, the “associated” user is passed as a parameter to the `deleteMe()` method, which simply deletes that user from the `users` array. The contents of `deleteMe()` is standard JavaScript code for removing an item from an array. You can replace the block of pseudocode in Listing 2.4 with the same code that you added in Listing 2.3 to prevent users from entering an empty string or a duplicate string.

Angular Directives and Child Components

In this section you will see how to create a child component in Angular that you can reference in an Angular application.

Copy the directory `ChildComponent` from the companion disc into a convenient location. Listing 2.5 displays the contents of `app.component.ts`

that illustrates how to import a custom component (written by you) in an Angular application.

LISTING 2.5: app.component.ts

```
import {Component}   from '@angular/core';

@Component({
   selector: 'app-root',
   template: `<div>Goodbye<child-comp></child-comp>World!
                                                   </div>`
})
export class AppComponent {}
```

Listing 2.5 contains a `template` property that consists of a `<div>` element that contains a nested `<child-comp>` element, where the latter is the value of the `selector` property in the child component `ChildComponent`.

Notice that Listing 2.5 does *not* import the `ChildComponent` class: this class is imported in `app.module.ts` in Listing 2.7.

Listing 2.6 displays the contents of `child.component.ts` in the `app` subdirectory.

LISTING 2.6: child.component.ts

```
import {Component} from '@angular/core';

@Component({
   selector: 'child-comp',
   template: `<div>Hello World from ChildComponent!</div>`
})
export class ChildComponent{}
```

Listing 2.6 is straightforward: The `template` property specifies a text string that will appear inside the `<child-comp>` element that is nested inside the `<div>` element in Listing 2.5.

NOTE *This is the first code sample in this chapter that involves modifying the default contents of the file* `app.module.ts`*.*

Listing 2.7 displays the modified contents of `app.module.ts`, which *must* import the class `ChildComponent` from `child.component.ts` and also specify `ChildComponent` in the `declarations` property. These additions are shown in bold.

LISTING 2.7: app.module.ts

```
import { NgModule }       from '@angular/core';
import { BrowserModule }  from '@angular/platform-browser';
import { AppComponent }   from './app.component';
import { ChildComponent } from './child.component';

@NgModule({
  imports:      [ BrowserModule ],
  declarations: [ AppComponent, ChildComponent ],
  bootstrap:    [ AppComponent ]
})
export class AppModule { }
```

The first detail to notice in Listing 2.7 is the new `import` statement (shown in bold) that imports the `ChildComponent` component from the TypeScript file `child.component.ts`. The second detail is the inclusion of `ChildComponent` (shown in bold) in the `declarations` array.

As you can see, these are fairly straightforward steps for including a child component in an Angular application. With practice you will become familiar with the sequence of steps that are illustrated in this section.

The Constructor and Storing State in Angular

This section contains a code sample that illustrates how to initialize a variable in a constructor and then reference the value of that variable via interpolation in the `template` property.

Copy the directory `StateComponent` from the companion disc into a convenient location. Listing 2.8 displays the contents of `app.component.ts`.

LISTING 2.8: app.component.ts

```
import {Component} from '@angular/core';

@Component({
  selector: 'app-root',
  template: '<h3>My name is {{emp.fname}} {{emp.lname}}</h3>'
})
export class AppComponent {
  public emp  = {fname:'John',lname:'Smith',city:'San
                                                    Francisco'};
```

```
    public name = 'John Smith'

    constructor() {
      this.name = 'Jane Edwards'
      this.emp  = {fname:'Sarah',lname:'Smith',city:'San
                                                      Francisco'};
    }
}
```

Listing 2.8 is almost the same as Listing 2.5: The current code involves the addition of a `constructor()` method that initializes the variable `name` as well as the literal object `emp`. The `emp` variable is shown in bold in the template property and also in two other places inside the `AppComponent` class.

Question: Which name will be displayed when you launch the application?

Answer: The value that is assigned to the `emp` variable in the `constructor`. This behavior is the same as OO-oriented languages such as Java.

Here is the output from launching this application:

My name is Sarah Smith

Keen-eyed readers will notice how we "slipped in" the TypeScript keyword `public` in the declaration of the `emp` and `name` variables. Other possible keywords include `private` and `protected`, which (again) behave the same way that they do in Java. If you are unfamiliar with these keywords, you can find online tutorials that will explain their purpose.

TypeScript supports another handy syntax (for variables), which is discussed in the next section.

Private Arguments in the Constructor: A Shortcut

TypeScript provides a short-hand notation for initializing private variables via a constructor. Consider the following TypeScript code block:

```
class MyStuff {
   private firstName: string;

   constructor(firstName: string) {
      this.firstName = firstName;
   }
}
```

A simpler and equivalent TypeScript code block is shown here:

```
class MyStuff {
   constructor(private firstName: string) {
   }
}
```

TypeScript support for the `private` keyword in a constructor is a nice feature: it reduces some boilerplate code and also eliminates a potential source of error (i.e., misspelled variable names).

As another example, the `constructor()` method in the following code snippet populates an `employees` object with data retrieved from an `EmpService` component (defined elsewhere and not important here):

```
constructor(private empService: EmpService) {
   this.employees = this.empService.getEmployees();
}
```

The next section shows you how to use the `*ngIf` directive for conditional logic in Angular applications.

Conditional Logic in Angular

Although previous examples contain a `template` property with a single line of text, Angular enables you to specify multiple lines of text. If you place interpolated variables inside a pair of matching backticks, Angular will replace ("interpolate") the variables with their values.

Copy the directory `IfLogic` from the companion disc into a convenient location. Listing 2.9 displays the contents of `app.component.ts` that illustrates how to use the `*ngIf` directive.

LISTING 2.9: app.component.ts

```
import {Component} from '@angular/core';

@Component({
  selector: 'app-root',
  template: `
     <h3>Hello everyone!</h3>
     <h3>My name is {{emp.fname}} {{emp.lname}}</h3>
     <button (click)="moreInfo()">More Details</button>
     <div *ngIf="showMore === true">
       <h3>I live in {{emp.city}}</h3>
     </div>
```

```
    <div (click)="showDiv = !showDiv">Toggle Me</div>
    <div *ngIf="showDiv"
         style="color:white;background-color:blue;
                                          width:25%">Content1</div>

    <div *ngIf="showDiv"
         style="background-color:red; width:25%;">Content2
                                                      </div>

})
export class AppComponent {
  public emp = {fname:'John',lname:'Smith',city:'San
                                                Francisco'};

  public showMore = false;

  moreInfo() {
    this.showMore = true;
  }
}
```

Listing 2.9 contains some new code in the `template` property: a `<button>` element that invokes the method `moreInfo()` when users click the button. After the click event, a `<div>` element with city-related information inside an `<h3>` element is displayed. Notice that this `<div>` element is only displayed when `showMore` is `true`, which is controlled via the `ngIf` directive that checks for the value of `showMore`. The initial value of `showMore` is `false`, but as soon as users click the `<button>` element, the value is set to `true`, and at that point the `<div>` element is displayed.

The new code in `AppComponent` involves a Boolean variable `showMore` (initially `false`) and the method `moreInfo()`, which initializes `show-More` to `true`.

Detecting Mouse Positions in Angular Applications

Angular provides support for detecting various mouse events, some of which you have already seen in previous sections (such as click events). The code sample in this section shows you how to detect a mouse position inside a Scalable Vector Graphics (SVG) `<svg>` element (SVG graphics effects are discussed in Chapter 3).

Copy the directory `SVGMouseMove` from the companion disc into a convenient location. Listing 2.10 displays the contents of `app.component.ts` that illustrates how to detect a `mousemove` event and to display the coordinates of the current mouse position.

LISTING 2.10: app.component.ts

```
import {Component} from '@angular/core';

@Component({
   selector: 'app-root',
   template: `<div><mouse-move></mouse-move></div>`
})
class AppComponent {}
```

Listing 2.10 contains a `template` property that consists of a `<div>` element that contains a nested `<mouse-move>` element, where the latter is the value of the `selector` property in the custom component `MouseMove`, which is defined in `mousemove.ts`. In essence, the component `AppComponent` "delegates" the handling of `mousemove` events to the `MouseMove` component, which defines the `mouseMove()` function in order to handle such events.

Listing 2.11 displays the contents of `mousemove.ts` that illustrates how to detect a `mousemove` event and to display the coordinates of the current mouse position.

LISTING 2.11: mousemove.ts

```
import {Component} from '@angular/core';

@Component({
 selector: 'mouse-move',
 template: `<svg id="svg" width="600px" height="400px"
                (mousemove)="mouseMove($event)">
           </svg>
           `
})
export class MouseMove{
   mouseMove(event) {
      console.log("Position x: "+event.clientX+" y: "+event.
                                                    clientY);
   }
}
```

Listing 2.11 contains the `mouseMove()` method whose lone argument `event` is an object that contains information (such as its location) about the mouse event. The `mouseMove()` method contains a `console.log()` statement that simply displays the x-coordinate and the y-coordinate of the location of the mouse click event.

Remember to update the contents of `app.module.ts` to include the `MouseMove` class, as shown in Listing 2.12.

LISTING 2.12: app.module.ts

```
import { NgModule }         from '@angular/core';
import { BrowserModule }  from '@angular/platform-browser';
import { AppComponent }   from './app.component';
import { MouseMove }        from './mousemove';

@NgModule({
  imports:       [ BrowserModule ],
  declarations: [ AppComponent, MouseMove ],
  bootstrap:     [ AppComponent ]
})
export class AppModule { }
```

Listing 2.12 imports the `MouseMove` class and adds this class to the `declarations` property (both of which are shown in bold).

Mouse Events and User Input in Angular

Angular provides support for mouse events, and automatically recognizes the events `click`, `mousedown`, `mousemove`, `mouseover`, `mouseup`, and `mousewheel`.

The following code snippet shows you how to specify a `<button>` element with an event handler in standard HTML:

```
<button onclick="action()">Action</button>
```

The following code snippet shows you how to specify a `<button>` element with an event handler in Angular:

```
<button (click)="action($event)">Action</button>
```

Listing 2.13 displays the contents of `mouseevents.ts` that illustrates how to handle a `mouseover` event in Angular.

LISTING 2.13: mouseevents.ts

```
import {Component} from '@angular/core';

@Component({
 selector: 'mouse-events',
 template: `<div>
              <input type="text" #myInput>
              <button (mouseover)="mouseEvent($event,myInput.
                                          value)">Mouse Over</button>
            </div>`
```

```
})
export class MouseEvents {
   mouseCount = 0;

   mouseEvent(event, value) {
      ++this.mouseCount;
      console.log("mouse count: "+this.mouseCount);
      console.log(event, value);
   }
}
```

Listing 2.13 contains a `template` property that comprises a `<div>` element, an `<input>` element where users can enter text, and a `<button>` element with a mouse-related event handler called `mouseEvent`. The expression `$event.myInput.value` references the text in the `<input>` element, and this value is passed to the `mouseEvent()` method.

The next portion of Listing 2.13 is the exported class `MouseEvents` that starts with the variable `mouseCount` whose initial value is `0`. The remainder of `MouseEvents` is the `mouseEvent()` method, which increments and displays the value of `mouseCount` during each `mouseover` event and displays the text in the `<input>` element.

Listing 2.14 displays the contents of `app.component.ts` that involves a "generic" `<mouse-events>` element for mouse-related events in Angular. Keep in mind that this element is a custom element (i.e., it's not an Angular element).

LISTING 2.14: app.component.ts

```
import {Component} from '@angular/core';

@Component({
   selector: 'app-root',
   template: `<div><mouse-events></mouse-events></div>`
})
export class AppComponent {}
```

Listing 2.14 contains a `template` property that consists of a `<div>` element that contains a nested `<mouse-events>` element, where the latter is the value of the `selector` property in the custom component `MouseEvents`, which is defined in `mouseevents.ts`. In essence, the component `AppComponent` "delegates" the handling of `mouseover` events to the `MouseEvents` component, which defines the `mouseEvent()` function in order to handle such events.

To capture user input via a `mouseclick`, replace `(mouseover)` with `(click)` in the `<button>` element (and also the displayed text), as shown here:

```
<button (click)="mouseEvent($event,myInput.value)">Add
                                                    </button>
```

Remember to update the contents of `app.module.ts` to include the `MouseEvents` class, as shown in Listing 2.15.

LISTING 2.15: app.module.ts

```
import { NgModule }        from '@angular/core';
import { BrowserModule } from '@angular/platform-browser';
import { AppComponent }   from './app.component';
import { MouseEvents }    from './mouseevents';

@NgModule({
  imports:       [ BrowserModule ],
  declarations: [ AppComponent, MouseEvents ],
  bootstrap:     [ AppComponent ]
})
export class AppModule { }
```

Listing 2.15 is straightforward: It imports the `MouseEvents` class and adds this class to the `declarations` property (shown in bold).

Handling User Input

The code sample in this section shows you how to handle user input and introduces the notion of a *service* in Angular, which is discussed in greater detail in Chapter 5.

As you have already seen, Angular enables you to create a reference to an HTML element, as shown here:

```
<input type="text" #user>
```

The `#user` syntax creates a reference to the `<input>` element that enables you to reference `{{user.value}}` to see its value, or `{{user.type}}` to see the type of the input. Moreover, you can use this reference in the following code block:

```
<p (click)="user.focus()">
  Get the input focus
```

```
</p>
<input type="text" #user (keyup)>
{{user.value}}
```

When users click the `<input>` element, the `focus()` method is invoked, and the `(keyup)` property updates the value in the input during the occurrence of a `keyup` event.

Copy the directory `TodoInput` from the companion disc into a convenient location. Listing 2.16 displays the contents of `app.component.ts` that illustrates how to reference a component that appends user input to an array in Angular.

LISTING 2.16: app.component.ts

```
import {Component}   from '@angular/core';

@Component({
   selector: 'app-root',
   template: `<div>
                <todo-input></todo-input>
                <todo-list></todo-list>
             </div>`
})
export class AppComponent {}
```

Listing 2.16 contains a standard `import` statement. The `template` property specifies a `<div>` element that contains placeholders for the `TodoInput` and `TodoList` components.

Listing 2.17 displays the contents of `todoinput.ts` that illustrates how to display an `<input>` field and a `<button>` element to capture user input in Angular.

LISTING 2.17: todoinput.ts

```
import {Component}    from '@angular/core';
import {TodoService } from './todoservice';

@Component({
 selector: 'todo-input',
 template: `
   <div>
     <input type="text" #myInput>
     <button (click)="mouseEvent(myInput.value)">Add Name
                                                  </button>
   </div>`
```

```
})
export class TodoInput{
   constructor(public todoService:TodoService) {}

   mouseEvent(value) {
      if((value != null) && (value.length > 0)) {
        this.todoService.todos.push(value);
        console.log("todos: "+this.todoService.todos);
      } else {
         console.log("value must be non-null");
      }
   }
}
```

Listing 2.17 contains a `template` property that consists of a `<div>` element that contains an `<input>` element for user input, followed by a `<button>` element for handling mouse click events.

The `TodoInput` class defines an empty constructor that also initializes an instance of the custom `TodoService` that is imported at the top of the file. This instance contains an array `todos` that is updated with new to-do items when users click the `<button>` element, provided that the new to-do item is not the empty string.

Listing 2.18 displays the contents of `todolist.ts` that keeps track of the items in a to-do list.

LISTING 2.18: todolist.ts

```
import {Component}    from '@angular/core';
import {TodoService} from './todoservice';

@Component({
 selector: 'todo-list',
 template: `<div>
           <ul>
             <li *ngFor="let todo of todoService.todos">
               {{todo}}
             </li>
           </ul>
          </div>`
})
export class TodoList {
   constructor(public todoService:TodoService) {}
}
```

Listing 2.18 contains a `template` property whose contents are similar to the contents of the `template` property in Listing 2.3, along with an empty constructor that initializes an instance of the `TodoService` custom component. This instance is used in the `template` property to iterate through the elements in the `todos` array.

Listing 2.19 displays the contents of `todoservice.ts` that keeps track of the to-do list.

LISTING 2.19: todoservice.ts

```
export class TodoService {
   todos = [];
}
```

Listing 2.19 contains a `todos` array that is updated with new to-do items when users click the `<button>` element in the root component.

Finally, update the contents of `app.module.ts` to include the class shown in bold in Listing 2.20.

LISTING 2.20: app.module.ts

```
import { NgModule }       from '@angular/core';
import { BrowserModule }  from '@angular/platform-browser';
import { AppComponent }   from './app.component';
import { TodoInput }      from './todoinput';
import { TodoList }       from './todolist';
import { TodoService }    from './todoservice';

@NgModule({
  imports:        [ BrowserModule ],
  providers:      [ TodoService ],
  declarations:   [ AppComponent, TodoInput, TodoList ],
  bootstrap:      [ AppComponent ]
})
export class AppModule { }
```

Listing 2.20 imports three to-do-related classes and adds them to the `providers` property and the `declarations` property (shown in bold).

The moduleId and templateUrl Properties in Angular

An earlier section showed you how to use the `ngFor` directive in a `template` property to iterate through a list of items. Angular also supports the `templateUrl` property, which means that you can move the code in the `template` property to a separate file, and then reference the name of that file as the value of the `templateUrl` property.

For example, suppose that the code in the `template` property is placed in the file `itemdetails.html`. If this file is in the same directory as `itemsapp.ts` (shown below), you must include the property `moduleId` to indicate the subdirectory where the file `itemdetails.html` is located. If you do not specify the `moduleId` property, then Angular assumes that `itemdetails.html` is in the same directory as the top-level Web page that launches the top-level Angular component.

NOTE *Use the `moduleId` property to specify relative paths for files that are specified in the `templateUrl` property.*

Listing 2.21 displays the contents of `itemsapp.ts` that references the file `itemdetails.html`, which contains the code for iterating through a list of items. The `moduleId` property indicates that this file is in the same directory as `itemsapp.ts`.

LISTING 2.21: itemsapp.ts

```
import {Component} from '@angular/core';

@Component({
  selector: 'app-root',
  moduleId: 'app/itemdetails',
  templateUrl: 'itemdetails.html'
})
export class AppComponent {
  items = [];

  constructor() {
    this.items = ['one','two','three','four'];
  }
}
```

Listing 2.21 contains a `templateUrl` property that references the file `itemdetails.html` with Angular code for displaying the contents of the items array.

Listing 2.22 displays the contents of `itemdetails.html` that contains the code for displaying the contents of the `items` array, which is initialized in the constructor in Listing 2.21.

LISTING 2.22: itemdetails.html

```
<div *ngFor="let item of items">
  {{item}}
</div>
```

You could easily move the contents of Listing 2.22 into a `template` property in Listing 2.21; the only reason for including this file is to show you how to use the `templateUrl` property.

Working with Custom Classes in Angular

You have already seen an example of a custom TypeScript class that represents a user. This section shows you how to work with an array of instances of a custom TypeScript class.

Listing 2.23 displays the contents of `newuser.ts` that illustrates how to create a custom TypeScript class that represents a user.

LISTING 2.23: newuser.ts

```
import {Component} from '@angular/core';

@Component({
   selector: 'new-user',
   template: '',
})
export class User {
   fname:string;

   constructor(fname:string) {
      this.fname = fname;
   }
}
```

For ease of illustration, Listing 2.23 defines a very simple `User` custom component that only keeps track of the `fname` property for a single user.

Listing 2.24 displays the contents of `app.component.ts` that uses the `User` custom component to populate an array with a set of users represented by instances of the `User` class and then display user-related information in a list. The `User` class is obviously more useful when you include other user-related properties in addition to the first name property.

LISTING 2.24: app.component.ts

```
import {Component} from '@angular/core';
import {User}       from './newuser';
```

```
@Component({
   selector: 'app-root',
   template: `
      <div>
        <input #fname>
        <button (click)="clickMe(fname.value)">Add User</
                                                        button>
        <ul>
          <li *ngFor="let user of users"
                                       (click)="onSelect(user)">
            {{user.fname}}
          </li>
        </ul>
      </div>`
})
export class AppComponent {
   newUser:User;

   users = [
             new User("Jane"),
             new User("Dave"),
             new User("Tom")
           ];

   clickMe(user) {
      console.log("creating new user: "+user);
      this.newUser = new User(user);
      this.users.push(this.newUser);
   }

   onSelect(user) {
      console.log("Selected user: "+JSON.stringify(user));
      var index = this.users.indexOf(user);
      this.users.splice(index,1);
   }
}
```

Listing 2.24 contains a `template` property that iterates through the list of `User` instances in the `users` array and displays the name contained in each instance. Notice how the `users` array is initialized in the `AppComponent` component: Three `User` instances are created from the `User` custom component that is defined in `newuser.ts`.

Listing 2.24 also contains a `clickMe()` method that is invoked when users click the `<button>` element, after which a new user is appended to the `users` array. Finally, Listing 2.24 contains an `onSelect()` method that is invoked when users select a different item in the list of users.

Click Events in Multiple Components

An Angular application can contain multiple components, each of which can declare event handlers with the same name. This section contains an example that shows you the order in which click events are handled in an Angular application.

Copy the directory `ClickItems` from the companion disc into a convenient location. Listing 2.25 displays the contents of `app.component.ts` that declares an `onClick()` event handler for each item in a list of items.

LISTING 2.25: app.component.ts

```
import {Component} from '@angular/core';
import {ClickItem} from './clickitem';

@Component({
   selector: 'app-root',
   styles:   [`li { display: inline; }`],
   template: `
    <div>
      <ul>
       <li><img (click)="onClick()"
             width="100" height="100" src="src/sample1.png">
                                                         </li>
       <li><img (click)="onClick()"
             width="100" height="100" src="src/sample2.png">
                                                         </li>
       <li><img (click)="onClick()"
             width="100" height="100" src="src/sample3.png">
                                                         </li>
      </ul>
    </div>
    `
})
export class AppComponent {
  onClick() {
    console.log("app.component.ts: you clicked me");
  }
}
```

The `template` property in Listing 2.25 displays an unordered list in which each item is a clickable PNG-based image. When users click one of the images, the `onClick()` method is invoked that simply displays a message via `console.log()`.

Listing 2.26 displays the contents of `clickitem.ts` that declares an `onClick()` event handler for each item in a list of items.

LISTING 2.26: clickitem.ts

```
import {Component} from '@angular/core';

@Component({
   selector: 'cclick',
   styleUrl: [` li { inline: block } `],
   template: `
     <div>
      <ul>
       <li><img (click)="onClick(100)"
                width="100" height="100" src="src/sample1.
                                                  png"></li>
       <li><img (click)="onClick(200)"
                width="100" height="100" src="src/sample2.
                                                  png"></li>
       <li><img (click)="onClick(300)"
                width="100" height="100" src="src/sample3.
                                                  png"></li>
      </ul>
     </div>
     `
})
export class ClickItem {
  onClick(id) {
    console.log("clickitem.ts clicked: "+id));
  }
}
```

Listing 2.26 is similar to Listing 2.25 in terms of functionality. Launch the application and click in various locations in your browser, and observe the different messages that are displayed in Chrome Web Inspector.

Working with `@Input`, `@Output`, and `EventEmitter`

Angular supports the `@Input` and `@Output` annotations to pass values between components. The `@Input` annotation is for variables that receive values from a parent component, whereas the `@Output` annotation sends (or "emits") data from a component to its parent component when the value of the given variable is modified.

The output from this code sample is anticlimactic. However, the purpose of this code sample is to draw your attention to some of the nonintuitive code snippets (especially in `app.module.ts`). Moreover, this code sample works correctly for version 2.1.5 of the TypeScript compiler, but it's possible that future versions will require modifications to the code (so keep this point in mind).

Now copy the directory `ParentChildEmitters` from the companion disc into a convenient location. Listing 2.27 displays the contents of `app.component.ts` that shows you how to update the value of a property of a child component from a parent component.

LISTING 2.27: app.component.ts

```
import {Component}      from '@angular/core';
import {EventEmitter}   from '@angular/core';
import {ChildComponent} from './childcomponent';

@Component({
  selector: 'app-root',
  providers: [ChildComponent],
  template: `
     <div>
      <child-comp [childValue]="parentValue"
         (childValueChange)="reportValueChange($event)">
      </child-comp>
     </div>
   `
})
export class AppComponent {
  public parentValue:number = 77;

  constructor() {
    console.log("constructor parentValue = "+this.
                                                parentValue);
  }

  reportValueChange(event) {
    console.log(event);
  }
}
```

The `template` property in Listing 2.27 has a top-level `<div>` element that contains a `<child-comp>` element that has two attributes, as shown here:

```
<child-comp [childValue]="parentValue"
            (childValueChange)="reportValueChange($event)">
</child-comp>
```

The `[childValue]` attribute assigns the value of `parentValue` to the value of `childValue`. Notice that the variable `parentValue` is defined in `AppComponent`, whereas the variable `childValue` is defined in `ChildComponent`. *This is how to pass a value from a parent component to a child component.*

Next, the `childValueChange` attribute is assigned the value that is returned from `ChildComponent` to the current ("parent") component. Keep in mind that the attribute `childValueChange` is updated only when the value of `childValue` (in the child component) is modified. *This is how to pass a value from a child component to a parent component.*

Keep in mind the following point: The child component *must* define a variable of type `EventEmitter` (such as `childValueChange`) to "emit" a modified value from the child component to the parent component.

The next portion of Listing 2.27 is a simple constructor, followed by the method `reportValueChange`, which contains a `console.log()` statement.

Listing 2.28 displays the contents of `childcomponent.ts` that shows you how to update the value of a property of a child component from a parent component.

LISTING 2.28: childcomponent.ts

```
import {Component}    from '@angular/core';
import {Input}        from '@angular/core';
import {Output}       from '@angular/core';
import {EventEmitter} from '@angular/core';

@Component({
  selector: 'child-comp',
  template: `
     <button (click)="decrement();">Subtract</button>
     <input type="text" [value]="childValue">
     <button (click)="increment();">Add</button>
  `
})
export class ChildComponent {
  @Input() childValue:number = 3;
  @Output() childValueChange = new EventEmitter();

  constructor() {
    console.log("constructor childValue = "+this.childValue);
  }
```

```
  increment() {
    this.childValue++;

    this.childValueChange.emit({
      value: this.childValue
    })
  }
  decrement() {
    this.childValue--;

    this.childValueChange.emit({
      value: this.childValue
    })
  }
}
```

Listing 2.28 contains a `template` property that has a "decrement" `<button>` element, an `<input>` field where users can enter a number, and also an "increment" `<button>` element. The first `<button>` element increments the value `<input>` field, whereas the second `<button>` element decrements the value.

The exported class `ChildComponent` contains the numeric variable `childValue`, which is decorated via `@Input()`, and whose value is set by the parent.

As you can see, the methods `increment()` and `decrement()` increase and decrease the value of `childValue`, respectively. In both cases, the modified value of `childValue` is then "emitted" back to the parent with this code block:

```
this.childValueChange.emit({
   value: this.childValue
})
```

Update the contents of `app.module.ts` as shown in Listing 2.29, *which is different from the code in previous examples in this chapter.*

LISTING 2.29: app.module.ts

```
import { NgModule }          from '@angular/core';
import {CUSTOM_ELEMENTS_SCHEMA} from '@angular/core';
import { BrowserModule }  from '@angular/platform-browser';
import { AppComponent }    from './app.component';
import { ChildComponent } from './childcomponent';

@NgModule({
  imports:       [ BrowserModule ],
```

```
  providers:     [ ChildComponent ],
  declarations: [ AppComponent ],
  bootstrap:     [ AppComponent ],
  schemas:       [CUSTOM_ELEMENTS_SCHEMA]
})
export class AppModule { }
```

If you specify `ChildComponent` in the `declarations` property instead of the `providers` property, you will probably see this error message:

```
"Can't bind to <child-comp> since it isn't a known native
property"
```

When you launch the Angular application in this section, the value that is displayed in the `<input>` element is `77`, which is the value in the parent component, and *not* the value that is assigned in the child component (which is `3`).

Presentational Components

Presentational components receive data as input and generate views as outputs (so they do not maintain the application state). Consider the following component:

```
@Component({
  selector: 'student-info',
  template: `<h2>{{studentDetails?.status}}</h2>
    <div class="container">
      <table class="table">
        <tbody>
        <tr *ngFor="let student of students">
            <td>{{student.fname}}</td>
            <td>{{student.lname}}</td>
        </tr>
        </tbody>
    </table>
</div>`
})
export class StudentDetailsComponent {
   @Input()
   studentDetails:StudentDetails;
}
```

The `StudentDetailsComponent` component has primarily presentational responsibilities. The component receives input data and displays that on the screen. As a result, this component is reusable.

By contrast, application-specific components (also called "smart" components) are tightly coupled to a specific Angular application. Thus, a smart component would have a presentation component (but not the converse).

Because data is passed to this component synchronously (not via an `Observable`), the data might not be present initially, which is the reason for including the so-called Elvis operator (the "?" in the template).

Styling Elements and Shadow DOM in Angular

Angular supports Cascading Style Sheets (CSS) encapsulation, which means that CSS selectors will only match elements that are defined in the same custom component. This encapsulation is available because of `ShadowDOM` emulation in Angular. In particular, this involves an `import` statement to import `ViewEncapsulation`, and also specifying one of the following values:

- ViewEncapsulation.Emulated
- ViewEncapsulations.Native
- ViewEncapsulation.None

The default for components is `ViewEncapsulation.Emulated`, which outputs namespaces of our class next to our styles and inherits global styles. By contrast, `ViewEncapsulations.Native` uses the Native `ShadowDOM` (which is not supported in all browsers), and loses global styles. Finally, `ViewEncapsulation.None` removes all style encapsulation in a component.

```
styles: [`
    #mydiv {
      font-size: 1rem;
      line-height: 1.25;
      color: #999;
      background-color: #ffcccc;
    }
    :global(body) {
      color: #666;
      background-color: #ccccff;
    }
  `],
```

Copy the directory `ViewEncapsulation` from the companion disc into a convenient location. Listing 2.30 displays the contents of app.component.ts that contains a CSS selector that matches the <button> element in app.component.ts but not the <button> element in index.html.

LISTING 2.30: app.component.ts

```
import {Component} from '@angular/core';
import {ViewEncapsulation} from '@angular/core';

@Component({
  selector: 'app-root',
  encapsulation: ViewEncapsulation.Native,
  styles: [`
    .button { background-color: red; }
  `],
  template: `
    <button class="button">Click in Component</button>
  `,
})
class AppComponent {}
```

Listing 2.30 contains two import statements and an encapsulation property whose value is set to ViewEncapsulation.Native. Next, the styles property specifies the color red for the <button> element that is declared in the template section.

Now modify the <body> element in index.html by adding a <button> element as shown here:

```
<body>
  <app-root>Loading...</app-root>
  <button class="button">Click in index.html</button>
</body>
```

Launch the Angular application and you will see a red <button> element in the component and a dark gray <button> element in the HTML page.

Angular UI Components

The code samples in this chapter show you how to use HTML widgets in Angular applications. However, you can also use Angular UI components that are "wrappers" for HTML widgets. The official Angular website contains an extensive collection of UI components:

https://angular.io/resources/#!#UI%20Components

The preceding website contains various links to other toolkits, such as `ng-bootstrap` (native Angular 2 directives for Bootstrap) and Angular Material 2 (Material Design components for Angular 2).

The following website contains an extensive collection of Angular UI components (e.g., table, tree, menu, and chart):

https://github.com/brillout/awesome-angular-components

Keep in mind that the UI components in the preceding link are for Angular 2, so it's a good idea to test them in the latest version of Angular. In addition, Chapter 5 contains form-related UI components that do work in version 4 of Angular.

You can also perform an Internet search for other open source (or commercial) alternatives to determine which one suits your needs.

New Features in Angular

Some new features in Angular include support for if/else logic in the `ngIf` directive and the `NgComponentOutlet` directive.

The `ngIf` directive conditionally includes a template based on the value of an expression. Next, `ngIf` evaluates the expression and then renders the `then` or `else` template in its place when expression is `truthy` or `falsy` respectively. Typically the `then` template is the inline template of `ngIf` unless bound to a different value, and the `else` template is blank unless it is bound. More information is located here:

https://angular.io/docs/ts/latest/api/common/index/NgIf-directive.html

The `NgComponentOutlet` directive is an experimental directive that provides a declarative approach for creating dynamic components, an example of which is shown here:

```
@Component({
  selector: 'app-root',
  template: `
    <h1>Angular ngComponentOutlet</h1>
    <ng-container *ngComponentOutlet="myComponent">
                                                   </ng-container>
    <button (click)="doSomething()">Toggle Component</button>
  `,
})
```

As you can probably surmise, the `ng-container` directive is a logical container for "grouping" nodes. In the earlier code block, the `ng-container` directive has an `NgComponentOutlet` whose value is of type `Input`, which in turn references a custom component. Next, make sure you add dynamic components to the `entryComponents` section of `ngModule`, as shown here:

```
@NgModule({
  ...,
  entryComponents: [MyComponent1, MyComponent2],
  ...
})
```

The `NgComponentOutlet` directive supports additional options that are described here:

> *https://angular.io/docs/ts/latest/api/common/index/NgComponentOutlet-directive.html*

Summary

This chapter showed you how to use UI Controls in Angular applications. You saw how to render buttons, how to render lists of names, and how to add and delete names from those lists. You also learned about conditional logic and how to create child components.

Then you saw how to handle mouse-related events, such as `mousemove` events. Next you learned about communicating between parent and child components, followed by a discussion of presentational components. You also learned how to specify different types of CSS encapsulation in an Angular application. Finally, you were briefly introduced to some new UI-related features in Angular.

CHAPTER 3

GRAPHICS AND ANIMATION

This chapter shows you how to create Angular applications with graphics and animation effects via HTML5 technologies and various open source toolkits. This graphics-related chapter is introduced early in this book because graphics provide an enjoyable approach to learning technologies. Even if you do not plan to use graphics and animation in your Angular applications, it's worth skimming the contents of this chapter to familiarize yourself with visual effects that might be useful to you in the future.

As you will soon see, the code samples in this chapter use technologies that are unrelated to Angular, such as Scalable Vector Graphics (SVG), Data-Driven Documents (D3), GreenSock Animation Platform (GSAP), Cascading Style Sheets 3 (CSS3), and jQuery. In fact, one sample uses "pure" CSS3 to create animation effects in an Angular application, which is then adapted to a code sample that uses the Angular `animations` property to create animation effects. After reading the code in that example, you can create your own variations, perhaps by a combination of techniques from the other code samples in this chapter.

If you are a novice regarding graphics effects, this is probably the quickest path to follow to learn how to create graphics and animation effects with Angular. Thus, you benefit for two reasons: First, animation effects "the Angular way" have a decent amount of online documentation; second, you won't need to spend a lot of extra time learning how to create animation effects in Angular with unrelated technologies.

The first section in this chapter discusses the Angular lifecycle, in part because one of the lifecycle methods is required in the GSAP-related code sample in this chapter.

The second section in this chapter shows you how to render graphics and create animation effects in SVG as well as D3. Although the examples in this section are very simple, they provide a starting point in case you need to create such effects in your Angular application.

The third section contains examples of D3-based graphics and animation effects. In case you don't already know, D3 is an open source toolkit that is extremely popular and well-suited for data visualization (hence its inclusion in this chapter).

The fourth section contains an example of creating animation effects using "pure" CSS3, and a code sample that combines CSS3 with jQuery. Next you will learn how to handle mouse-related events in Angular applications. The final section briefly discusses the Angular module `ng2-charts` for creating charts and graphs.

An important caveat about the code samples in this chapter: They assume that you have a basic knowledge of SVG, D3, GSAP, CSS3, and jQuery. If you are unfamiliar with any of these technologies, you can read online tutorials that describe their basic features, or forge ahead in the code samples to identify the concepts that you need to learn from online sources. Even if you decide to skip the graphics samples, please read the first section of this chapter, which discusses the Angular lifecycle and provides useful information regarding Angular applications.

Various GitHub repositories are included that contain a plethora of swatch-like code samples that illustrate techniques for creating additional graphics and animation effects via SVG, D3, HTML5 Canvas, GSAP, and CSS3. As a side point (in case you are interested), it's also possible to combine WebGL with Angular, an example of which is located here:

https://github.com/chliebel/angular2-3d-demo

There is one other point to keep in mind: Try to avoid Document Object Model (DOM) operations with native solutions, such as `document.domMethod()` or `$('dom-element')`, and use them sparingly (if at all). Angular enables you to perform DOM operations safely via `ElementRef`, `Renderer` and `ViewContainer` application programming interfaces

(APIs). However, some code samples in this chapter do break the preceding recommendation to show you how to create a specific type of graphics effect.

NOTE

When you copy a project directory from the companion disc, if the node_modules directory is not present, then copy the top-level node_modules directory that has been soft-linked inside that project directory (which is true for most of the sample applications).

Before delving into the graphics-related code samples, let's look at the Angular lifecycle methods.

Angular Lifecycle Methods

Angular applications have lifecycle methods where you can place custom code to handle various events (application start, run, and so forth). The Lifecycle Hook interfaces are defined in the `@angular/core` library, and they are listed here:

- OnInit
- OnDestroy
- DoCheck
- OnChanges
- AfterContentInit
- AfterContentChecked
- AfterViewInit
- AfterViewChecked

Each interface has a single method whose name is the interface name prefixed with `ng`. For example, the `OnInit` interface has a method named `ngOnInit`. Angular invokes these lifecycle methods in the following order:

- ngOnChanges: called when an input or output binding value changes
- ngOnInit: after the first ngOnChanges
- ngDoCheck: developer's custom change detection
- ngAfterContentInit: after component content initialized
- ngAfterContentChecked: after every check of component content
- ngAfterViewInit: after component's view(s) are initialized
- ngAfterViewChecked: after every check of a component's view(s)
- ngOnDestroy: just before the directive is destroyed.

Because Angular invokes the constructor of a component when that component is created, the constructor is a convenient location to initialize the state for that component. However, child components must be initialized before accessing any properties or data that are defined in those child components. In this scenario, place custom code in the `ngOnInit` lifecycle method to access data from child components.

The complete set of Angular lifecycle events is located here:

https://angular.io/docs/ts/latest/guide/lifecycle-hooks.html

http://learnangular2.com/lifecycle/

A Simple Example of Angular Lifecycle Methods

Copy the directory `LifeCycle` from the companion disc into a convenient location.

Listing 3.1 displays the contents of `app.component.ts` that shows you the sequence in which some Angular lifecycle methods are invoked.

LISTING 3.1: app.component.ts

```
import {Component} from '@angular/core';

@Component({
  selector: 'app-root',
  template: '<h2>Angular Lifecycle Methods</h2>',
})
export class AppComponent{
   ngOnInit() {
     // invoked after child components are initialized
     console.log("ngOnInit");
   }
   ngOnDestroy() {
     // invoked when a component is destroyed
     console.log("ngOnDestroy");
   }
   ngDoCheck() {
     // custom change detection
     console.log("ngDoCheck");
   }
   ngOnChanges(changes) {
     console.log("ngOnChanges");
     // Invoked after bindings have been checked
```

```
    // but only if one of the bindings has changed.
    //
    // changes is an object of the format:
    // {
    //   'prop': PropertyUpdate
    // }
  }
  ngAfterContentInit() {
    // Component content has been initialized
    console.log("ngAfterContentInit");
  }
  ngAfterContentChecked() {
    // Component content has been checked
    console.log("ngAfterContentChecked");
  }
  ngAfterViewInit() {
    // Component views are initialized
    console.log("ngAfterViewInit");
  }
  ngAfterViewChecked() {
    // Component views have been checked
    console.log("ngAfterViewChecked");
  }
}
```

Listing 3.1 contains all the Angular lifecycle methods, where each method contains `console.log()` so that you can see the order in which the methods are executed.

Launch the application by navigating to the `src` subdirectory of the `LifeCycle` application, and invoke the following command:

```
ng serve
```

Navigate to `localhost:4200` in a Chrome session, and then open Chrome Web Inspector and you will see the following output in the `Console` tab:

```
ngOnInit
ngDoCheck
ngAfterContentInit
ngAfterContentChecked
ngAfterViewInit
ngAfterViewChecked
ngDoCheck
ngAfterContentChecked
ngAfterViewChecked
```

The next section illustrates the usefulness of the `ngAfterContent-Init()` method to apply animation effects after dynamically generating a set of SVG elements.

GSAP Animation and the `ngAfterContentInit()` Method

Copy the directory `D3GSAP` from the companion disc into a convenient location. This code sample involves updating (or creating) the following files:

package.json
main.ts
app.component.ts
app.module.ts
ArchTubeOvals1.ts

Now install the required `gsap` package in `package.json` as follows:

```
npm install gsap --save
```

Next, import `gsap` in `main.ts`, as shown in Listing 3.2.

LISTING 3.2: main.ts

```
import { enableProdMode } from '@angular/core';
import { platformBrowserDynamic }
        from '@angular/platform-browser-dynamic';

import { AppModule } from './app/app.module';
import { environment } from './environments/environment';
import 'gsap';

if (environment.production) {
  enableProdMode();
}

platformBrowserDynamic().bootstrapModule(AppModule);
```

Except for the code snippet in bold, Listing 3.2 contains dynamically generated code.

Now let's look at Listing 3.3, which displays the contents of `app.component.ts` that contains the lifecycle method `ngAfterContentInit()`.

This method executes GSAP-based animation code to animate some dynamically generated SVG elements.

LISTING 3.3: app.component.ts

```
import { Component } from '@angular/core';
import { TweenMax }  from 'gsap';
import 'gsap';

@Component({
   selector: 'app-root',
   template: '<svg></svg>'
})
export class AppComponent {
   ngAfterContentInit() {
     var deltaAngle = 1, maxAngle = 721;

     for(var angle=0; angle<maxAngle; angle+=deltaAngle) {
        var index = Math.floor(angle/deltaAngle);

        if(index % 3 == 0) {
          TweenMax.to("#elem"+angle, 3,
            {rotation:180, transformOrigin:"left top"});
          TweenMax.to("#elem"+angle, 8,
            {scale:1.5, rotationX:45, rotationY:225,
             x:10, y:0, z:200});
        } else if(index % 3 == 1) {
          TweenMax.fromTo("#elem"+angle, 2, {x: '+=400px'},
            {x: 100, y:50,
             scaleX:0.4, scaleY:0.4});
        } else {
          TweenMax.fromTo("#elem"+angle, 4, {x: '+=400px'},
            {x: 50, y:50,
             scaleX:0.3, scaleY:0.3});
        }
     }
   }
}
```

The `ngAfterContentInit` method in Listing 3.3 contains basic GSAP-based code for creating animation effects. These effects are applied to SVG elements based on the value of their `id` attribute, which is a concatenation of the string `elem` and an integer. Navigate to the GSAP home page and consult the online GSAP documentation for details regarding the GSAP APIs in Listing 3.3.

The `ngAfterContentInit()` method is useful when you need to execute a JavaScript function after a component has been loaded. In this code sample, the TypeScript file `ArchTubeOvals1.ts` defines a child component that dynamically generates a set of SVG elements, after which the code in `main-ArchOvals.ts` executes a JavaScript function that adds `GSAP`-based animation effects. This execution sequence ensures that the animation effects are applied *after* the SVG elements have been created: otherwise the SVG elements do not exist yet and so there is no animation effect!

Listing 3.4 displays the contents of `ArchTubeOvals1.ts` (in the `src/app` subdirectory) that dynamically generates a set of SVG elements.

LISTING 3.4: ArchTubeOvals1.ts

```
import {Component} from '@angular/core';

@Component({
 selector: 'svg',
 template: ''
})
export class ArchTubeOvals1 {
   constructor() {
      this.generateGraphics();
   }

   generateGraphics() {
     var svgns = "http://www.w3.org/2000/svg";
     var svg   = document.getElementById("svg");
     var colors = ["#ff0000", "#0000ff"];

     var basePointX   = 240,  basePointY   = 200;
     var currentX     = 0,    currentY     = 0;
     var offsetX      = 0,    offsetY      = 0;
     var majorX       = 30,   majorY       = 50;
     var Constant     = 0.25, angle        = 0;
     var deltaAngle   = 1,    maxAngle     = 721;
     var radius       = 1;

     for(angle=0; angle<maxAngle; angle+=deltaAngle) {
        radius   = Constant*angle;
        offsetX  = radius*Math.cos(angle*Math.PI/180);
        offsetY  = radius*Math.sin(angle*Math.PI/180);
        currentX = basePointX+offsetX;
        currentY = basePointY-offsetY;

        var ellipse = document.createElementNS(svgns,
                                                     "ellipse");
        ellipse.setAttribute("id", "elem"+angle);

        ellipse.setAttribute("cx", ""+currentX);
        ellipse.setAttribute("cy", ""+currentY);
```

```
        ellipse.setAttribute("rx", ""+majorX);
        ellipse.setAttribute("ry", ""+majorY);

        ellipse.setAttribute("fill", colors[angle % colors.
                                                        length]);
        svg.appendChild(ellipse);
      }
    }
}
```

Listing 3.4 defines the `ArchTubeOvals1` custom component whose constructor invokes the `generateGraphics()` method, which generates a set of SVG `<ellipse>` elements. The code in this method consists of a set of JavaScript variables followed by a loop that calculates positions on an Archimedean-like spiral. Each position is used to compute attributes for a dynamically generated SVG `<ellipse>` element that is rendered at that location.

Now update `app.module.ts` to include the necessary references to the TypeScript file `ArchTubeOvals1.ts`, as shown in bold in Listing 3.5.

LISTING 3.5: app.module.ts

```
import { BrowserModule }  from '@angular/platform-browser';
import { NgModule }       from '@angular/core';
import { FormsModule }    from '@angular/forms';
import { HttpModule }     from '@angular/http';
import { AppComponent }   from './app.component';
import { ArchTubeOvals1 } from './ArchTubeOvals1';

@NgModule({
  declarations: [ AppComponent, ArchTubeOvals1 ],
  imports: [
    BrowserModule,
    FormsModule,
    HttpModule
  ],
  providers: [],
  bootstrap: [AppComponent]
})
export class AppModule { }
```

The code sample in this section is a simple illustration of the graphics effects that you can create by combining Angular with `GSAP`. Various samples involving SVG, `GSAP`, and Angular (but with beta-level Angular code) are located here:

https://github.com/ocampesato/react-svg-gsapi

The next section shows you how to add CSS3 animation effects in Angular applications.

CSS3 Animation Effects in Angular

The code sample in this section enhances the code sample in the previous section by adding a CSS3 animation effect.

Copy the directory `SimpleCSS3Anim` from the companion disc into a convenient location. Listing 3.6 displays the contents of `app.component.ts` that illustrates how to change the color of list items when users hover over each list item with their mouse.

LISTING 3.6: app.component.ts

```
import {Component} from '@angular/core';

@Component({
   selector: 'app-root',
   template: '
     <h2>Employee Information</h2>
     <ul>
       <li *ngFor="let emp of employees">
         {{emp.fname}} {{emp.lname}} lives in {{emp.city}}
       </li>
     </ul>
    ',
    styles: ['
      @keyframes hoveritem {
          0%   {background-color: red;}
          25%  {background-color: #880;}
          50%  {background-color: #ccf;}
          100% {background-color: #f0f;}
      }

      li:hover {
          width: 50%;
          animation-name: hoveritem;
          animation-duration: 4s;
      }
    ']
})
export class AppComponent {
  employees = [];

  constructor() {
    this.employees = [
     {"fname":"Jane","lname":"Jones","city":"San Francisco"},
```

```
      {"fname":"John","lname":"Smith","city":"New York"},
      {"fname":"Dave","lname":"Stone","city":"Seattle"},
      {"fname":"Sara","lname":"Edson","city":"Chicago"}
    ];
  }
}
```

Listing 3.6 contains the `styles` property, which contains a `@keyframes` definition for creating an animation effect involving color changes. The `styles` property also contains an `li:hover` selector that references the `@keyframes` definition and specifies a time duration of `4` seconds for the animation effect. The colors that you see are specified in the `@keyframes` definition.

Launch the Angular application. When the list of names is displayed in a browser, move your mouse slowly over each name and watch how it changes color. Although this example is admittedly quite simple, you can modify its contents to achieve other CSS3-based animation effects.

Animation Effects via the "Angular Way"

The code sample in this section also creates an animation effect by adding some CSS3 selectors.

Copy the directory `SimpleNG2Anim` from the companion disc into a convenient location. Listing 3.7 displays the contents of `app.component.ts` that illustrates how to move the position of the `<li>` elements when users hover over them with their mouse. This code is based on modifications to the code in Listing 3.6 (as discussed later).

LISTING 3.7: app.component.ts

```
// part #1: new import statement
import {
  Component,
  Input,
  trigger,
  state,
  style,
  transition,
  animate
} from '@angular/core';

// part #2: new Emp class
class Emp {
  constructor(public fname: string,
```

```
              public lname: string,
              public city:  string,
              public state = 'inactive') {
  }

  toggleState() {
    this.state = (this.state==='active' ? 'inactive' :
                                                  'active');
    console.log(this.fname+" "+"new state = "+this.state);
  }
}
@Component({
   selector: 'app-root',
   // part #3: new animations property
   animations: [
     trigger('empState', [
       state('inactive', style({
         backgroundColor: '#eee',
         transform: 'scale(1)'
       })),
       state('active',   style({
         backgroundColor: '#cfd8dc',
         transform: 'scale(1.1)'
       })),
       transition('inactive => active', animate('100ms
                                                  ease-in')),
       transition('active => inactive', animate('100ms
                                                  ease-out'))
     ])
   ],
   template: `
     <h2>Employee Information</h2>
     <ul>
       <li *ngFor="let emp of employees"
                   [@empState]="emp.state"
                   (mousemove)="emp.toggleState()">
         {{emp.fname}} {{emp.lname}} lives in {{emp.city}}
       </li>
     </ul>
   `
})
export class AppComponent {
  employees = [];

  constructor() {
    // part #5: array of Emp objects
    this.employees = [
     new Emp("Jane","Jones","San Francisco"),
     new Emp("John","Smith","New York"),
     new Emp("Dave","Stone","Seattle"),
```

```
      new Emp("Sara","Edson","Chicago")
    ];
  }
}
```

Listing 3.7 consists of five modifications to the code in Listing 3.6. Specifically, the section labeled "part #1" is a new `import` statement that replaces the original `import` statement. The section labeled "part #2" is the newly added `Emp` class, which holds data for each employee.

The section labeled "part #3" is the new `transitions` property, which defines the behavior when an animation event is triggered (this occurs during a `mousemove` event "over" an `<li>` element). The portion in bold (which is not labeled but is "part #4") in the `ngFor` element essentially binds the `mousemove` event to the `toggleState()` method in the `Emp` class. Finally, the section labeled "part #5" is an array of `Emp` objects that replaces the original array in which each employee is represented as a JavaScript Object Notation (JSON) string.

A Basic SVG Example in Angular

The code sample in this section shows you how to specify a custom component that contains SVG code for displaying an SVG element. This example serves as the foundation for the code in the next section, which involves dynamically creating and appending an SVG element to the DOM.

Copy the directory `SVGEllipse1` from the companion disc into a convenient location. Listing 3.8 displays the contents of `app.component.ts` that references an Angular custom component in order to render an SVG ellipse.

LISTING 3.8: app.component.ts

```
import {Component} from '@angular/core';

@Component({
   selector: 'app-root',
   template: '<div><my-svg></my-svg></div>'
})
export class AppComponent {}
```

Listing 3.8 is very simple: the code defines a component whose `template` property contains a custom `<my-svg>` element inside a `<div>` element.

Listing 3.9 displays the contents of `MyEllipse1.ts` that contains the SVG code for an SVG ellipse.

LISTING 3.9: MyEllipse1.ts

```
import {Component} from '@angular/core';

@Component({
   selector: 'my-svg',
   template: `
     <svg width="500" height="300">
       <ellipse cx="100" cy="100"
                rx="50" ry="30"
                fill="red"/>
     </svg>
     `
})
export class MyEllipse1{}
```

Listing 3.9 is straightforward: The `template` property contains an SVG `<svg>` element with `width` and `height` attributes, and a nested SVG `<ellipse>` element with hard-coded values for the required attributes `cx`, `cy`, `rx`, `ry`, and `fill`.

Listing 3.10 displays the contents of `app.module.ts` with the new contents shown in bold.

LISTING 3.10: app.module.ts

```
import {Component}          from '@angular/core';
import { NgModule }         from '@angular/core';
import { BrowserModule }    from '@angular/platform-browser';
import { AppComponent }     from './app.component';
import { MyEllipse1 }       from './MyEllipse1';

@NgModule({
  imports:      [ BrowserModule ],
  declarations: [ AppComponent, MyEllipse1 ],
  bootstrap:    [ AppComponent ]
})
export class AppModule { }
```

Listing 3.10 contains generic code that you are familiar with from Chapter 2, as well as a new `import` statement (shown in bold) involving the

MyEllipse1 class. The other modification in Listing 3.10 is the inclusion of the MyEllipse1 class (shown in bold) in the declarations array.

Launch the Angular application and you will see a colored SVG ellipse.

Incidentally, the following links explain how to create SVG gradients and then how to create SVG Gradient Effects in Angular applications:

https://developer.mozilla.org/en-US/docs/Web/SVG/Tutorial/Gradients

https://medium.com/@OlegVaraksin/how-to-proper-use-svg-gradients-in-angularjs-2-3241672e4de2#.oah0e9z1k

Angular and Follow-the-Mouse in SVG

The code sample in this section creates a child component and uses mouse-related events to create dynamic graphics effects. Copy the directory SVGFollowMe from the companion disc into a convenient location.

Listing 3.11 displays the contents of app.component.ts that illustrates how to reference a custom Angular component that renders an SVG <ellipse> element at the current mouse position.

LISTING 3.11: app.component.ts

```
import {Component} from '@angular/core';

@Component({
   selector: 'app-root',
   template: '<div><mouse-move></mouse-move></div>'
})
export class AppComponent {}
```

As you can see, the template property in Listing 3.11 specifies a <div> element that contains a custom <mouse-move> element.

Listing 3.12 displays the contents of MouseMove.ts that illustrates how to reference a custom Angular component that renders an SVG <ellipse> element at the current mouse position.

LISTING 3.12: MouseMove.ts

```
import {Component} from '@angular/core';

@Component({
```

```
  selector: 'mouse-move',
  template: `<svg id="svg" width="600" height="400"
              (mousemove)="mouseMove($event)">
            </svg>
            `
})
export class MouseMove {
   radiusX = "25";
   radiusY = "50";

   mouseMove(event) {
     var svgns = "http://www.w3.org/2000/svg";
     var svg   = document.getElementById("svg");
     var colors = ["#ff0000", "#88ff00", "#3333ff"];

     var sum = Math.floor(event.clientX+event.clientY);

     var ellipse = document.createElementNS(svgns, "ellipse");
     ellipse.setAttribute("cx", event.clientX);
     ellipse.setAttribute("cy", event.clientY);
     ellipse.setAttribute("rx", this.radiusX);
     ellipse.setAttribute("ry", this.radiusY);
     ellipse.setAttribute("fill", colors[sum % colors.length]);
     svg.appendChild(ellipse);
   }
}
```

Listing 3.12 contains a `template` property that defines an SVG `<svg>` element. The `(mousemove)` event handler is executed whenever users move their mouse, which in turn executes the custom method `mouseMove()`.

Notice that the `mouseMove` method accepts an `event` argument, which is an object that provides the coordinates of the location of each `mousemove` event. The coordinates are specified by `event.clientX` and `event.clientY`, which are the x-coordinate and the y-coordinate, respectively, of the current mouse position.

The next code block in the `mouseMove` method dynamically creates an SVG `<ellipse>` method, sets the values of the five required attributes (see the previous section for the details), and then appends the newly created SVG `<ellipse>` method to the DOM. This functionality creates a follow-the-mouse effect that you can see when you launch the Angular application code in this section.

Note that the final line of code in the `mouseMove` method appends an SVG `<ellipse>` element *directly* to the DOM; it is better to avoid this approach if it's possible to do so.

Listing 3.13 displays the contents of `app.module.ts` with the new contents shown in bold.

LISTING 3.13: app.module.ts

```
import { BrowserModule } from '@angular/platform-browser';
import { NgModule }      from '@angular/core';
import { FormsModule }   from '@angular/forms';
import { HttpModule }    from '@angular/http';
import { AppComponent }  from './app.component';
import { MouseMove }     from './MouseMove';

@NgModule({
  declarations: [ AppComponent, MouseMove ],
  imports: [
    BrowserModule,
    FormsModule,
    HttpModule
  ],
  providers: [],
  bootstrap: [AppComponent]
})
export class AppModule { }
```

The code in Listing 3.13 follows a familiar pattern: Starting with the "baseline" code, add an `import` statement that references an exported TypeScript class (which is `MouseMove` in this example) and add that same TypeScript class to the `declarations` array.

Launch the Angular application. After a new browser session is launched, slowly move your mouse to see the different colored SVG ellipses rendered near your mouse. As an exercise, modify the code in `MouseMove.ts` so that new SVG ellipses are "centered" underneath your mouse.

D3 and Angular

The previous two sections showed you examples of Angular applications with SVG. This section shows you how to combine D3 with Angular. Note that the code sample in this section also appends SVG elements directly to the DOM.

In case you don't already know, D3 is an open source toolkit that provides a JavaScript-based layer of abstraction over SVG. Fortunately, the attributes of every SVG element are the same attributes that you specify in D3 (so your work is cut in half).

Copy the directory D3Angular2 from the companion disc into a convenient location. Listing 3.14 displays the contents of app.component.ts that illustrates how to use D3 to render basic SVG graphics in an Angular application.

LISTING 3.14: app.component.ts

```
import { Component, ViewChild, ElementRef } from '@angular/
                                                       core';
import * as d3 from 'd3';

//------------------------------------
// Keep in mind the following points
// when you want to use d3 in Angular:
// 1) npm install d3 --save
// 2) import * as d3 from 'd3'
// 3) note the <div> in 'template'
// 4) the ViewChild(...)  code snippet
// 5) the "nativeElement" code snippet
//------------------------------------

@Component({
   selector: 'app-root',
   template: '<div id="mysvg" #mysvg></div>'
})
export class AppComponent {
  @ViewChild('mysvg') mysvg: ElementRef;

  constructor() {}

  ngAfterContentInit() {
     this.createSVG();
  }

  //-----------------------------------------------------
  // view children are only set when ngAfterViewInit()
  // is invoked and content children are only set when
  // ngAfterContentInit() is invoked.
  //
  // Since the method createSVG() is invoked after the
  // ngAfterContentInit() method, the <div> in the
  // template property is available (i.e., non-null).
  //-----------------------------------------------------

  createSVG() {
     var width = 1000, height = 800;

     // circle and ellipse attributes
     var cx = 50,  cy = 80, radius1 = 40,
         ex = 250, ey = 80, radius2 = 80;
```

```
    // color/rectangle/line segment attributes
    var colors = ["red", "blue", "green"];
    var rectX  = 15,  rectY = 200;
    var rWidth = 100, rHeight = 40;
    var x1=170,y1=200,x2=320,y2=200,lineWidth=8;

    let svgElement = this.mysvg.nativeElement;

    // create an SVG element
    let svg = d3.select(svgElement)
                .append("svg")
                .attr("width",  width)
                .attr("height", height);

    // append a circle
    svg.append("circle")
       .attr("cx", cx)
       .attr("cy", cy)
       .attr("r",  radius1)
       .attr("fill", colors[0]);

    // append an ellipse
    svg.append("ellipse")
       .attr("cx", ex)
       .attr("cy", ey)
       .attr("rx", radius2)
       .attr("ry", radius1)
       .attr("fill", colors[1]);

    // append a rectangle
    svg.append("rect")
       .attr("x", rectX)
       .attr("y", rectY)
       .attr("width",  rWidth)
       .attr("height", rHeight)
       .attr("fill", colors[2]);

    // append a line segment
    svg.append("line")
       .attr("x1", x1)
       .attr("y1", y1)
       .attr("x2", x2)
       .attr("y2", y2)
       .attr("stroke-width", lineWidth)
       .attr("stroke", colors[0]);
  }
}
```

Listing 3.14 starts with two `import` statements, followed by a comment block that summarizes the key points for using D3.js in Angular applications. The `template` property contains a `<div>` element that is available in the ngAfterContentInit method, which in turn simply invokes the createSVG() method, which populates an SVG <svg> element with four shapes (a circle, an ellipse, a rectangle, and a line segment).

Note the @ViewChild decorator that defines the variable mysvg that has type ElementRef; this variable "links" the <div> element in the template property with the variable svgElement, which is defined in the createSVG() method:

```
let svgElement = this.mysvg.nativeElement;
```

Notice how the various SVG elements are dynamically created and how their mandatory attributes (which depend on the SVG element in question) are assigned values via the attr() method, as shown here (and in the preceding code block as well):

```
// append a circle
svg.append("circle")
   .attr("cx", cx)
   .attr("cy", cy)
   .attr("r",  radius1)
   .attr("fill", colors[0]);
```

After you learn the mandatory attribute names for SVG elements, you can use the preceding syntax to create and append such elements to the DOM.

You can also find many similar code samples involving SVG and Angular (with beta-version Angular code) here:

https://github.com/ocampesato/angular2-svg-graphics

D3 Animation and Angular

The following code block illustrates how to add D3-based animation effects to the SVG <circle> element in the D3Angular2 Angular application:

```
svg.on("mousemove", function() {
  index = (++moveCount) % circleColors.length;

  var circle = svg.append("circle")
                  .attr("cx", (width-100)*Math.random())
                  .attr("cy", (height-100)*Math.random())
                  .attr("r",  radius)
                  .attr("fill", circleColors[index])
                  .transition()
```

```
                    .duration(duration)
                    .attr("transform", function() {
                        return "scale(0.5, 0.5)";
                      //return "rotate(-20)";
                    })
});
```

The code inside the preceding event handler is executed during each `mousemove` event, accompanied by the dynamic creation of an SVG `<ellipse>` element. The new functionality involves the `transition()` method, the `duration()` method, and setting the `transform` attribute, all of which are shown in bold in the preceding code block.

As you can see, the `transform` attribute is set to a `scale()` value, which sets the width and height to `50%` of their initial value during an interval of `2` seconds (which equals `2000` milliseconds), thereby creating an animation effect.

Pure CSS3 3D Animation in Angular

The code sample in this section creates 3D graphics and animation effects without any SVG, D3, HTML5 `Canvas` code, or any other graphics-related toolkit.

Copy the directory `PureCSS3Anim` from the companion disc into a convenient location. Listing 3.15 displays the contents of `index.html` that references a CSS selector with 3D animation effects.

LISTING 3.15: index.html

```
<!DOCTYPE html>
<html>
 <head>
  <meta charset="utf-8">
  <title>Angular and CSS3 Animation</title>
  <base href="/">

  <meta name="viewport" content="width=device-width,
                                          initial-scale=1">
  <link rel="icon" type="image/x-icon" href="favicon.ico">
  <link rel="stylesheet" type="text/css"
        href="Anim240Flicker3DLGrad2SkewOpacity2Reflect1DIV6.
                                                         css">
 </head>

 <body>
```

```
      <app-root><div id="mysvg">Loading...</div></app-root>
      <div id="outer">
       <div id="linear1">Text1</div>
       <div id="linear2">Text2</div>
       <div id="linear3">Text3</div>
       <div id="linear4">Text4</div>
       <div id="linear5">Text5</div>
       <div id="linear6">Text6</div>
      </div>
  </body>
</html>
```

Listing 3.15 contains the usual Angular code, followed by a `<body>` element that contains a `<div>` with six nested `<div>` elements. Each of these `<div>` elements matches a CSS selector that creates an animation effect.

Listing 3.16 displays a small portion of the contents of the CSS stylesheet `Anim240Flicker3DLGrad2SkewOpacity2Reflect1DIV6.css`.

LISTING 3.16: Anim240Flicker3DLGrad2SkewOpacity2Reflect1DIV6.css

```
#outer {
  position: relative; top: 10px; left: 0px;
}

@-webkit-keyframes upperLeft {
   0% {
      -webkit-transform: matrix(1.5, 0.5,  0.0, 1.5, 0, 0)
                         matrix(1.0, 0.0,  1.0, 1.0, 0, 0);
   }
   10% {
      -webkit-transform: translate3d(50px,50px,50px)
rotate3d(30,40,50,-90deg) skew(-15deg,0) scale3d(1.25, 1.25,
                                                       1.25);
   }
   20% {
      -webkit-transform: matrix(1.0, 1.5, -0.5, 1.0, 0, 0)
                         matrix(0.5, 0.5,  0.5, 0.5, 0, 0);
   }
   25% {
      -webkit-transform: matrix(0.4, 0.5,  0.5, 0.3, 250, 50)
                         matrix(0.3, 0.5, -0.5, 0.4, 50, 150);
   }
   30% {
      -webkit-transform: perspective(200px)
rotate3d(20,30,40,-180deg) skew(105deg,0) scale3d(1.25, 1.25,
                                                       1.25);
   }
// details omitted for brevity
```

```
    98% {
       -webkit-transform: matrix(0.4, 0.5,  0.5, 0.3, 200, 50)
                          matrix(0.3, 0.5, -0.5, 0.4, 50, 150);
    }
    99% {
       -webkit-transform: translate3d(150px,50px,50px)
rotate3d(6,8,10, 240deg) skew(315deg,0) scale3d(1.0, 0.7,
                                                         0.3);
    }
    100% {
       -webkit-transform: matrix(1.0, 0.0,  0.0, 1.0, 0, 0)
                          matrix(1.0, 0.5,  1.0, 1.5, 0, 0);
    }
}
```

Listing 3.16 contains a small portion of the code in the CSS stylesheet called `Anim240Flicker3DLGrad2SkewOpacity2Reflect1DIV6`.css. Consult online documentation and tutorials that contain details regarding CSS3 `@keyframes`.

CSS3 and jQuery Animation Effects in Angular

The code sample in this section contains some CSS3 gradient effects whose details are beyond the scope of this sample. However, if you intend to create this type of gradient effect, you can find online tutorials that provide background details.

Copy the directory `CSS3JQueryAnim` from the companion disc into a convenient location. Listing 3.17 displays the contents of `index.html` that contains JavaScript code for creating graphics effects based on jQuery and CSS3.

LISTING 3.17: index.html

```
<!DOCTYPE html>
<html>
  <head>
   <meta charset="utf-8">
   <title>Angular and Archimedean Ellipses</title>

   <link href="SkewAnim1.css" rel="stylesheet" type="text/css">
   <script src="http://code.jquery.com/jquery-1.7.1.min.js">
   </script>

   <style>
```

```
      #outer {
        position: absolute;
        width: 90%; height: 90%;
        border: solid 2px #000;
      }

      .radial6 {
        background-color:white;
        background-image:

        -webkit-radial-gradient(red 4px, transparent 18px),
        -webkit-repeating-linear-gradient(
                        45deg, red 0px,  green 4px,
                        yellow 8px, blue 12px,
                        transparent 28px, green 20px, red 24px,
                        transparent 28px, transparent 32px),
        -webkit-repeating-linear-gradient(
                        -45deg, red 0px,  green 4px,
                        yellow 8px, blue 12px,
                        transparent 28px, green 20px, red 24px,
                        transparent 28px, transparent 32px);

        background-size: 50px 60px, 70px 80px;
        background-position: 0 0;
        -webkit-box-shadow:  30px 30px 30px #000;
        resize:both;
        overflow:auto;
      }
    </style>
  </head>

  <body>
    <app-root>Loading...</app-root>
    <div id="outer">
    </div>
  </body>
</html>
```

The `<head>` tag in Listing 3.17 contains a `<link>` tag that references the CSS stylesheet `SkewAnim1.css`, which contains the animation-related code. The next code snippet is a reference to jQuery code (which you can replace with a later version).

The remainder of the `<head>` tag is a `<style>` element that contains two selectors. The first selector matches an HTML element (in `index.html`) whose `id` attribute has the value `outer`. The second selector matches elements whose `class` attribute has the value `radial6`. Where are those elements? As you will soon see, they are programmatically generated (with some help from jQuery) in a loop in Listing 3.18.

Listing 3.18 displays the contents of app.component.ts that contains JavaScript code for creating graphics and animation effects based on jQuery and CSS3.

LISTING 3.18: app.component.ts

```
import { Component } from '@angular/core';

declare var $:any;

@Component({
  selector: 'app-root',
  template: ''
})
export class AppComponent {
  constructor() {
    $(document).ready(function() {
      var fillRed    = "rgb(255, 0, 0)";
      var fillYellow = "rgb(255, 255, 0)";
      var fillColor  = "rgb(255, 0, 0)";

      var basePointX = 300,  basePointY = 150;
      var majorAxis  = 40,   minorAxis  = 80;
      var currentX   = 0,    currentY   = 0;
      var offsetX    = 0,    offsetY    = 0;
      var deltaAngle = 3,    maxAngle   = 720;
      var Constant   = 0.25, radius     = 0;
      var newNode;

      for(var angle=0; angle<maxAngle; angle++) {
         radius   = Constant*angle;
         offsetX  = radius*Math.cos(angle*Math.PI/180);
         offsetY  = radius*Math.sin(angle*Math.PI/180);
         currentX = basePointX+offsetX;
         currentY = basePointY-offsetY;

         if(Math.floor(angle/deltaAngle) % 2 == 0) {
            fillColor = fillRed;
         } else{
            fillColor = fillYellow;
         }

         // create an ellipse at the current position
         if(angle % 20 == 0) {
           newNode = $('<div>').css({'position':'absolute',
                                   'width':majorAxis+'px',
                                  'height':minorAxis+'px',
                                    left: currentX+'px',
                                    top: currentY+'px',
```

```
                                    'backgroundColor': fillColor,
                                     'borderRadius': '20%'
                                  }).
                               toggleClass("skewAnim1");
            } else {
              newNode = $('<div>').css({
                           'position':'absolute',
                           'width':majorAxis+'px',
                           'height':minorAxis+'px',
                           left: currentX+'px',
                           top: currentY+'px',
                          //'backgroundSize': '40px 40px, 180px
                                                          180px',
                            'backgroundSize': '240px 240px, 80px
                                                           80px',
                            'backgroundColor': fillColor,
                            'borderRadius': '50%'
                        }).
                        addClass("radial6 glow");
            }

            $("#outer").append(newNode);
        }
      });
    }
}
```

Listing 3.19 contains a standard `import` statement, followed by this code snippet:

```
declare var $: any;
```

The preceding snippet is necessary for TypeScript to "find" jQuery, which is loaded via a `<script>` element in `index.html`.

Listing 3.19 exports the TypeScript class `AppComponent` whose constructor contains all the code for dynamically generating HTML `<div>` elements and then appending them to the DOM.

The first part of the constructor starts by initializing some JavaScript variables for creating graphics. Next, a standard jQuery "ready" code snippet is included, which guarantees that the code inside this snippet is executed after the DOM has been loaded into memory. Specifically, there is a loop that calculates positions that approximately follow an Archimedean spiral, after which the jQuery `css()` method dynamically creates and appends `<div>` elements at those locations to the DOM. The loop contains simple if-else conditional logic to specify values for different properties. During each iteration, the following code snippet appends the newly created `<div>` element to the DOM element whose `id` attribute has the value `outer`:

```
$("#outer").append(newNode);
```

Listing 3.19 displays a portion of the contents of the CSS stylesheet `SkewAnim.css` that contains the CSS selectors for creating animation effects.

LISTING 3.19: SkewAnim1.css

```
@-webkit-keyframes glow {
  0% {
    -webkit-box-shadow: 0 0 24px rgba(255, 255, 255, 0.5);
  }
  50% {
    -webkit-box-shadow: 0 0 24px rgba(255, 0, 0, 0.9);
  }
  100% {
    -webkit-box-shadow: 0 0 24px rgba(255, 255, 255, 0.5);
  }
}

.skewAnim1 {
  -webkit-transform : skew(60deg, -20deg) scale(0.75, 1.75)
                                                  rotate(-60deg);
  -transform : skew(60deg, -20deg) scale(0.75, 1.75)
                                                  rotate(-60deg);
  -webkit-box-shadow: 8px 8px 8px #f00;
  box-shadow: 8px 8px 8px #f00;
  -webkit-animation-name: animCube1;
  -webkit-animation-duration: 10s;
}

.skewAnim2 {
  -webkit-transform : skew(-60deg, 50deg) scale(1.5, 0.75)
                                                  rotate(140deg);
  transform : skew(-60deg, 50deg) scale(1.5, 0.75)
                                                  rotate(140deg);
  -webkit-box-shadow: 8px 8px 8px #f00;
  box-shadow: 8px 8px 8px #f00;
  -webkit-animation-name: animCube1;
  -webkit-animation-duration: 10s;
}
// details omitted for brevity
```

Listing 3.19 contains a CSS3 `@keyframes` property (with vendor-specific prefixes) that creates a complex visual effect. An example of such a CSS selector containing a `transform` property that invokes the `skew()` function, the `scale()` function, and the `rotate()` function is shown here:

```
.skewAnim1 {
  -webkit-transform : skew(60deg, -20deg) scale(0.75, 1.75)
                                                    rotate(-60deg);
  transform : skew(60deg, -20deg) scale(0.75, 1.75)
                                                    rotate(-60deg);
  -webkit-box-shadow: 8px 8px 8px #f00;
  box-shadow: 8px 8px 8px #f00;
  -webkit-animation-name: animCube1;
  -webkit-animation-duration: 10s;
}
```

The preceding code block produces a transformation effect that involves skewing, scaling, and rotational transformations, along with shadow effects. The two lines shown in bold in the preceding code block "link". the animation effects in **`animCube1`** (defined elsewhere) to the selector `skewAnim1`, and specify that the duration of the animation effect is 10 seconds (10s).

Animation Effects "the Angular Way"

Now that you have seen Angular applications that create graphics and animation effects using various other technologies, this section contains a code sample with animation effects that involves CSS-based functionality and a small amount of Bootstrap code. If you are unfamiliar with Bootstrap, you can still follow the rest of the code and view the animation effects.

Copy the directory `NgGraphicsAnimation` from the companion disc into a convenient location. Listing 3.20 displays the contents of `app.component.html` that contains HTML markup for rendering two `<div>` elements and two `<button>` elements that trigger animation effects.

LISTING 3.20: app.component.html

```
<div class="container">
  <div class="row">
    <div class="col-xs-12">
      <h1>Angular Animation Effects</h1>
      <button class="btn btn-primary"
              (click)="onAnimate()">Animate Elements</button>
      <hr>
```

```
      <div style="width: 200px; height: 100px"
           [@divState]="state"
           (@divState.start)="animBegin($event)"
           (@divState.done)="animComplete($event)">
      </div>
      <br>
      <div style="width: 200px; height: 100px"
           [@currState]="currState">
      </div>
    </div>
  </div>
  <hr>
</div>
```

Listing 3.20 contains three nested `<div>` elements with a Bootstrap `container` class, `row` class, and `col-xs-12` class respectively. The innermost `<div>` element contains a `<button>` element that triggers the animation effects (shown in Listing 3.21) when users click the button. Clicking the button invokes the method `onAnimate()`, which updates the value of the variables `state` and `currState`.

The next portion of Listing 3.20 is a `<div>` element that is updated based on the value of the `state` property, which can be either `normal` or `animated`. Each of these two values has a corresponding entry in the `animations` property that you will see in Listing 3.21. This `<div>` element is displayed as an ellipse because the `width` and `height` properties are different and because of the `border-radius` property.

Notice that the second `<div>` element has a `@currState` property that is based on the value of `currState`, whereas the first `<div>` element is based on the value of `state`.

Listing 3.21 displays the contents of `app.component.ts` that contains the Angular `animations` property. This property contains code that transforms and animates `<div>` elements via CSS-based animation effects.

LISTING 3.21: app.component.ts

```
import { Component, trigger, state, style, transition,
            animate, keyframes, group } from '@angular/core';

// more details regarding browseranimationmodule:
// http://stackoverflow.com/questions/43362898/whats-
the-difference-between-browseranimationsmodule-and-
noopanimationsmodule
```

```
//----------------------------------------
// make sure you perform the following step:
// npm install @angular/animations --save
//----------------------------------------

@Component({
  selector: 'app-root',
  templateUrl: './app.component.html',
  animations: [
    trigger('divState', [
      state('normal', style({
            'background-color': '#008888',
            borderRadius: '50%',
            transform: 'translateX(0)'
      })),
      state('animated', style({
            'background-color': 'blue',
            transform: 'rotate3d(50,50,50,-180deg) skew
(-15deg,0) scale3d(1.25, 1.25, 1.25)'
      })),
      transition('normal <=> animated', animate(500)),
    ]),
    trigger('currState', [
      state('normal', style({
            'background-color': 'red',
            transform: 'translateX(0) scale(1)'
      })),
      state('animated', style({
            'background-color': 'green',
            transform: 'translateX(300px) scale(0.5)'
      })),
      transition('normal   => animated', animate(500)),
      transition('animated => normal', animate(1500)),
      transition('animated <=> *', [
        style({
          'background-color': '#880000'
        }),
        animate(2000, style({
          borderRadius: '50px'
        })),
        animate(500)
      ])
    ]),
  ]
})
export class AppComponent {
  state     = 'normal';
  currState = 'normal';

  onAnimate() {
```

```
    this.state == 'normal' ? this.state = 'animated' : this.
                                              state = 'normal';
    this.currState == 'normal' ? this.currState = 'animated' :
                                      this.currState = 'normal';
  }

  animBegin(event) {
    console.log(event);
  }

  animComplete(event) {
    console.log(event);
  }
}
```

Listing 3.21 contains a lengthy block of code for the animations property, which consists of two trigger functions (both are shown in bold). The first trigger executes a block of code based on whether the value of `divState` is `normal` or `animated`. Similarly, the second trigger executes a block of code based on whether the value of `currState` is `normal` or `animated`.

In all four cases, a simple set of CSS properties are updated to set the background color and the transform method. In addition to the CSS transforms in this code sample, you can use many other CSS transforms, including `rotate()`, `perspective()`, `matrix()`, and other 3D CSS3 transforms.

The interesting transform value is in the second state of the first trigger, as shown here:

```
transform: 'rotate3d(50,50,50,-180deg) skew(-15deg,0)
                                   scale3d(1.25, 1.25, 1.25)'
```

The preceding `transform` is a combination of a 3D rotation, a 2D skew effect, and a 3D scale effect. If you need to create complex visual effects, be assured that CSS3 provides an incredibly powerful set of transforms for creating rich and aesthetically appealing visual effects. The choice of visual effects obviously depends on the target audience (e.g., corporate environment versus high school students).

Notice that the second `trigger()` function contains several `transition()` functions, which specify the behavior of the `<div>` elements. For example, the first `transition()` specifies a duration of `500` milliseconds when making the transition from `normal` to `animated`, whereas the opposite transition occurs during `1500` milliseconds. The third transition() specifies a duration of 500 milliseconds during the update of the background-color property and the borderRadius property.

Listing 3.22 displays the contents of app.module.ts, with the new contents shown in bold.

LISTING 3.22: app.module.ts

```
import { BrowserModule } from '@angular/platform-browser';
import { BrowserAnimationsModule }
         from '@angular/platform-browser/animations';
import { NgModule }        from '@angular/core';
import { FormsModule }     from '@angular/forms';
import { HttpModule }      from '@angular/http';
import { AppComponent }    from './app.component';

@NgModule({
  declarations: [
    AppComponent
  ],
  imports: [
    BrowserAnimationsModule,
    BrowserModule,
    FormsModule,
    HttpModule
  ],
  providers: [],
  bootstrap: [AppComponent]
})
export class AppModule { }
```

The code in Listing 3.22 contains code that is familiar to you, along with code that is shown in bold, which is necessary for the animation-related effects.

Chart Tools for Angular

There are several open source toolkits available that provide chart-related functionality. One of them is the Angular module ng2-charts for creating charts and graphs, and its home page is located here:

http://mean.expert/2016/09/17/angular-2-chart-component-revised/

Another option is the `ng2d3` framework, which is an Angular2 + D3js composable reusable charting framework whose home page is located here:

https://github.com/swimlane/ng2d3

The `ng2d3` framework uses Angular to render and animate the SVG elements, and D3 for the math functions, scales, axis and shape generators, and so forth. Note that Angular does the actual rendering. In addition, `ng2d3` supports custom charts, and styles are customizable through CSS.

Check the supported features in these (and other) toolkits to determine which one best suits your needs.

Summary

This chapter showed you how to render SVG-based graphics in an Angular application. You learned how to create graphics and animation effects with D3 and GSAP, and in the latter case, you saw how to place custom code in one of the Angular lifecycle methods so that the animation effects are applied after the SVG elements have been generated. In addition, you learned how to use pure CSS3 graphics and animation effects in Angular.

CHAPTER 4

HTTP REQUESTS AND ROUTING

This chapter shows you how to make HTTP requests in Angular applications and how to work with Observables in Angular applications. The code samples show you how to obtain data from various sources, and how to create an `Observable` from a `Promise`. If need be, you can access various online tutorials containing introductory material about Observables and Promises.

The first section briefly discusses Dependency Injection (DI) and the `@Injectable` decorator. The second section shows you how to make an HTTP request in an Angular application to read JavaScript Object Notation (JSON)-based data defined in a text file.

The third section shows you how to make an HTTP request in an Angular application to retrieve information about a GitHub user. You will also see how to make multiple concurrent requests via the `forkJoin()` method. The final section in this chapter discusses routing in Angular applications.

NOTE

When you copy a project directory from the companion disc, if the node_modules directory is not present, then copy the top-level node_modules directory that has been soft-linked inside that project directory (which is true for most of the sample applications).

Dependency Injection in Angular

Angular provides a simple mechanism for dependency injection: a dependency is injected into the constructor of a class. You can inject multiple dependencies by specifying each dependency as an argument in a constructor of a class.

DI involves specifying the `@Injectable` decorator above a TypeScript class, and a constructor with a type that you want to be injected. The code samples in this chapter use the `Http` service, which is also imported in TypeScript files.

There are two simple steps that you need to perform: First, import `Http` in `app.component.ts` (and possibly other custom classes) and then update the contents of `app.module.ts`.

For example, the following code block (which is Step 1) injects an instance of the `Http` class (shown in bold):

```
import {Injectable} from '@angular/core';
import {Http} from '@angular/http';
...
@Injectable()
export class AppComponent {
  constructor(http:Http) {
     this.http = http;
  }
}
```

NOTE *The `Http` module uses `rxjs` to return `Observables` in Angular.*

As you can see in the preceding code block, the `Injectable` service is in `@angular/core`, whereas the `Http` service is in `@angular/http`.

NOTE *You must also update the contents of `app.module.ts` when you import `Http`.*

Step 2 involves updating `app.module.ts`, as shown here:

```
import { NgModule }        from '@angular/core';
import { BrowserModule } from '@angular/platform-browser';
import { HttpModule }     from '@angular/http';

@NgModule({
  imports:       [ BrowserModule, HttpModule ],
  declarations: [ AppComponent ],
  bootstrap:     [ AppComponent ]
})
export class AppModule { }
```

Notice that the preceding code block imports `HttpModule` from `@angular/http` and specifies it as a dependency in the `imports` property, whereas Step 1 imports `Http` from `@angular/http`. Later in this chapter, you will see code samples that require the `providers` property in `app.module.ts`.

Flickr Image Search Using jQuery and Angular

The code sample in this section shows you how to use jQuery in an Angular application, which is relevant for existing Web pages that perform `HTTP` requests via jQuery.

Copy the directory `SearchFlickr` from the companion disc into a convenient location. Note that the file `index.html` contains the following code snippet, which enables the use of jQuery in this project:

```
<script src="http://code.jquery.com/jquery-latest.js"> </script>
```

Listing 4.1 displays the contents of `app.component.ts` that illustrates how to make an `HTTP GET` request to retrieve images from `Flickr`. This request is based on text string that users enter in a search box.

LISTING 4.1: app.component.ts

```
import {Component} from '@angular/core';
declare var $: any;

@Component({
   selector: 'app-root',
   template: `
       Enter a word and search for related images:
       <br />
       <input id="searchterm" />
       <button (click)="httpRequest()">Search</button>
       <div id="images"></div>
   `
})
export class AppComponent {
  url = "http://api.flickr.com/services/feeds/photos_public.
                                          gne?jsoncallback=?";

  constructor() {}

  httpRequest() {
    $.getJSON(this.url,
    {
      tags: $("#searchterm").val(),
      tagmode: "any",
```

```
          format: "json"
        },
        function(data) {
          $.each(data.items, function(i,item){
            $("<img/>").attr("src", item.media.m).
                                       prependTo("#images");
          });
        });
      }
    }
```

Listing 4.1 contains a standard `import` statement, followed by this code snippet:

```
declare var $: any;
```

The preceding snippet is necessary for TypeScript to "find" jQuery, which is loaded via a `<script>` element in `index.html`. If you remove the preceding code snippet, you will see the following error:

```
app/app.component.ts(20,5): error TS2304: Cannot find name '$'.
```

The next portion of Listing 4.1 is the `@Component` decorator, whose `template` property contains `<input>`, `<button>`, and `<div>` elements to capture the user's search string, perform a search with that string, and display the results of the search, respectively.

The next portion of Listing 4.1 is the exported class `@AppComponent` that defines the `url` variable that is initialized with a hard-coded string value that "points" to the Flickr website.

Next, an empty constructor is defined, followed by the `httpRequest()` method that is invoked when users click the `<button>` element. This method invokes the jQuery `getJSON()` method that performs a Flickr image search based on the text string entered in the `<input>` element because of this code snippet:

```
tags: $("#searchterm").val()
```

When the matching images are retrieved, they are available via `data.items`, and the jQuery `each()` method iterates through the list of images. Each image is dynamically inserted in the `<images>` element via this snippet:

```
$("<img/>").attr("src", item.media.m).prependTo("#images");
```

FIGURE 4.1 A partial list of images showing pasta.

Take a minute or two to absorb the compact manner in which jQuery achieves the desired result.

Figure 4.1 displays the output from launching this Angular application and searching Flickr with the keyword `pasta`.

Combining Promises and Observables in Angular

The code sample in this section shows you how to retrieve `JSON`-based data, convert that data into a `Promise`, and then convert the `Promise` into an `Observable`. In addition, you will see two techniques for handling the data: One code block handles the data as a `Promise`, and another code block handles the data as an `Observable` (so you have a choice of either style). Remember that the `Http` service in Angular returns an `Observable` that supports a `subscribe()` method.

Navigate to the following website to sign up for the free application programming interface (API) key that you will need for the code sample in this section:

http://developer.nytimes.com/signup

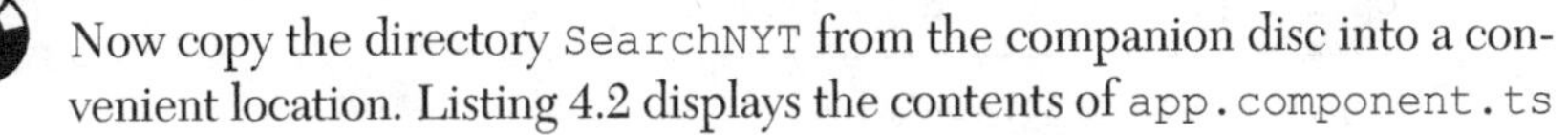

Now copy the directory `SearchNYT` from the companion disc into a convenient location. Listing 4.2 displays the contents of `app.component.ts`

that illustrates how to convert a `Promise` (that is returned from a custom service) into an `Observable` and then display the returned data as a list of links.

LISTING 4.2: app.component.ts

```
import {Component}      from '@angular/core';
import {Observable}     from 'rxjs/Observable';
import 'rxjs/Rx';
import {NYTService}     from './nyt-service';
declare var $: any;

@Component({
  selector: 'app-root',
  template: `
    <div>
      <h2>New York Times Headlines For Today</h2>
      <ul>
        <li *ngFor="let item of headlines">
          <a href="#">{{item.headline.main}}"</a>
        </li>
      </ul>
    </div>
  `
})
export class AppComponent {
  headlines: any;

  constructor(private nytService:NYTService) {
     var p = new Promise(function(resolve) {
        var value = nytService.getNYTInfo();
        resolve(value);
     });
//----------------------------------------
// Option #1: process data from an Observable
//----------------------------------------
     Observable.fromPromise(p)
       .subscribe(
         data => this.headlines = data,
         err => console.log('error reading data: '+err),
         () => this.documentInfo()
       );

/*
//----------------------------------------
// Option #2: process data from a Promise
//----------------------------------------
     p.then(function(data) {
```

```
        this.headlines = data.response.docs;
        console.log("found headlines = " +
                    JSON.stringify(this.headlines));
    }
*/
  }
  documentInfo() {
    this.headlines = this.headlines.response.docs;
  }
}
```

Listing 4.2 contains an `@Component` decorator whose `template` property displays an unordered list of headlines from articles that are retrieved from the *New York Times* website.

The `AppComponent` class contains three parts. The first part defines a `Promise p` that invokes the `getNYTInfo()` method (defined in the `NYTService` class shown below) to obtain a set of articles. The second part defines an `Observable` from the `Promise p`, whose `subscribe()` method initializes the variable `headlines` with the list of headlines from the retrieved articles via this code snippet:

```
data => this.headlines = data.response.docs
```

The `*ngFor` statement in the `template` iterates through the items in `headlines` to display the list of article headlines.

The third part obtains the list of articles from the `Promise p`, and initializes the variable `headlines` with the list of headlines from the retrieved articles.

Listing 4.3 displays the contents of `nyt.service.ts` that uses jQuery to make an `HTTP GET` request from a *New York Times* endpoint, which then returns a `Promise` to the parent component. Note that the endpoint will return a `JSON` string, and that the `return` statement (shown in bold in Listing 4.3) converts that `JSON` string into a `Promise`.

LISTING 4.3: nyt.service.ts

```
import {Inject, Injectable} from '@angular/core';
import {Http}               from '@angular/http';
declare var $: any;

@Injectable()
export class NYTService {
  // register for your NYT account and then get an API key:
  apiKey = "c96f0026207946e8ab610f4c7abcxxxxxxxxx";
```

```
  nytURL = "http://api.nytimes.com/svc/search/v2/
                                         articlesearch.json";

  constructor(@Inject(Http) public http:Http) { }

  getNYTInfo() {
    return $.get(this.nytURL, {
      "api-key": this.apiKey,
      sort:"oldest",
      fq:"headline:(\""+"Fashion"+"\")",
      fl:"headline,snippet,multimedia,pub_date"}, function(res) {
        var responseObj =
             $.parseJSON(JSON.stringify(res.response));

        var docs = JSON.stringify(responseObj["docs"]);
        var docsArray = JSON.parse(docs);
        this.dataFound = true;
        return docsArray;
      }, "JSON")
  }
}
```

Listing 4.3 initializes the variables `apiKey` and `nytURL` with the value of the registered API key and the *New York Times* URL, respectively. The main focus is the `gettNYTInfo()` method, which that populates various attributes (such as `fq` and fl) that are required by the *New York Times*, and then invokes the jQuery `get()` method to retrieve articles from the *New York Times*.

Here's the interesting part: Because of the `return` statement (shown in bold), *the JSON-based data is returned as a* `Promise`. The `docsArray` contains an array of articles that is accessed in the code in Listing 4.2.

Listing 4.4 displays the updated contents of `app.module.ts` that imports the `NYTService` class and the Angular `HttpModule`.

LISTING 4.4: app.module.ts

```
import { NgModule }       from '@angular/core';
import { BrowserModule } from '@angular/platform-browser';
import { HttpModule }     from '@angular/http';
import { AppComponent }   from './app.component';
import { NYTService }     from './nyt.service';

@NgModule({
  imports:      [ BrowserModule, HttpModule ],
  providers:    [ NYTService ],
  declarations: [ AppComponent ],
```

```
  bootstrap:     [ AppComponent ]
})
export class AppModule { }
```

The main thing to notice in Listing 4.4 is that the custom `NYTService` class is listed in the array of `providers`, whereas the Angular `HttpModule` is listed in the array of `imports`, both of which are specified in the `@NgModule` decorator.

Note that the Web service is less than 100% reliable, and you might see an error message similar to this one:

```
ERROR in SearchNYT/src/app/app.component.ts (36,39): Property
                        'response' does not exist on type '{}'.)
webpack: Failed to compile.
```

Wait a short while and reload the Web page and eventually you will see the correct results.

Figure 4.2 displays the output from launching this Angular application and retrieving a set of articles from the *New York Times* (admittedly the user interface [UI] portion can be greatly improved).

New York Times Headlines For Today

- Loss of the Brig Fashion, of Baltimore."
- MISSISSIPPI.; The Montgomery House at Pass Christian--A Southern Watering Place--Beauty and Fashion--Sea-bathing and Fine Fruit--R. H. Montgomery--The late Judge Preston--Dan Webster--The Scott men in the South, &c."
- Fashion and Sickness."
- GREAT BRITAIN.; DEATH OF THE DUKE OF WELLINGTON, The Duke's Public Career--Opinions--Macaulay--Reciprocity between France and England--Cholera--AEronautics--Currency--Literary Intelligence--The Court--Fashion--Theatricals, &c."
- Fashion and Dress; From the London Lady's Newspaper."
- Fashion and Dress."
- ORIGINAL JOTTINGS.; The Republican Fashion."
- WILLIAMSBURG CITY.; FOLLOWING THE FASHION. HOSPITAL REPORT. CAMPHENE ACCIDENT. DISHONEST SERVANTS."
- NEWPORT.; THE SEASON AT NEWPORT. Fashion, Festivities, and Miscellaneous Movements."
- " Fashion and Famine."--Letter form the Anthor."

FIGURE 4.2 A list of article headings from the *New York Times*.

Reading JSON Data via an Observable in Angular

This section shows you how to read data from a file that contains JSON-based data. Copy the directory `ReadJSONFile` from the companion disc into a convenient location. Listing 4.5 displays the contents of `app.component.ts` that illustrates how to make an `HTTP` request (which returns an `Observable`) to read a `JSON` -based file with `employee` information.

LISTING 4.5: app.component.ts

```
import { Component }  from '@angular/core';
import { Observable } from 'rxjs/Observable';
import { Inject }     from '@angular/core';
import { Http }       from '@angular/http';
import 'rxjs/Rx';
declare var $: any;

@Component({
  selector: 'app-root',
  template: `
     <button (click)="httpRequest()">Employee Info</button>
     <ul>
       <li *ngFor="let emp of employees">
         {{emp.fname}} {{emp.lname}} lives in {{emp.city}}
       </li>
     </ul>
  `
})
export class AppComponent {
  employees = [];

  constructor(@Inject(Http) public http:Http) {}

  httpRequest() {
    this.http.get('src/app/employees.json')
      .map(res => res.json())
      .subscribe(
        // this function runs on success
        data => this.employees = data,
        // this function runs on error
        err => console.log('error reading data: '+err),
        // this function runs on completion
        () => this.userInfo()
      );
  }

  userInfo() {
 //console.log("employees = "+JSON.stringify(this.employees));
  }
}
```

The template property in Listing 4.5 starts with a `<button>` element for making an `HTTP GET` request to retrieve information about employees.

The `template` property also contains a `<ul>` element for displaying an unordered list of employee-based data.

The `AppComponent` class contains the variable `employees`, followed by a constructor that initializes the `http` variable, which is an instance of the `Http` class. The `httpRequest()` method contains the code for making the `HTTP GET` request that returns an `Observable`. The `subscribe()` method contains the usual code, which in this case also initializes the `employees` array from the contents of the file `employees.json` in the subdirectory `app/src`.

Listing 4.6 displays the contents of `employees.json` that contains employee-related information. This file is located in the `src/app` subdirectory, as shown in bold in Listing 4.5.

LISTING 4.6: employees.json

```
[
{"fname":"Jane","lname":"Jones","city":"San Francisco"},
{"fname":"John","lname":"Smith","city":"New York"},
{"fname":"Dave","lname":"Stone","city":"Seattle"},
{"fname":"Sara","lname":"Edson","city":"Chicago"}
]
```

Listing 4.7 displays the contents of `app.module.ts` that imports the Angular `HttpModule`.

LISTING 4.7: app.module.ts

```
import { NgModule }       from '@angular/core';
import { BrowserModule } from '@angular/platform-browser';
import { HttpModule }     from '@angular/http';
import { AppComponent }  from './app.component';

@NgModule({
  imports:      [ BrowserModule, HttpModule ],
  declarations: [ AppComponent ],
  bootstrap:    [ AppComponent ]
})
export class AppModule { }
```

Listing 4.7 contains the standard set of import statements, along with `HttpModule`, which is listed in the array of `imports` in the `@NgModule` decorator.

Employee Info

- Jane Jones lives in San Francisco
- John Smith lives in New York
- Dave Stone lives in Seattle
- Sara Edson lives in Chicago

FIGURE 4.3 A list of employees from a JSON file.

Figure 4.3 displays the output from launching this Angular application and retrieving JSON-based employee data from a file.

The next section contains an example of using `forkjoin()`, and the code is similar to the contents of Listing 4.5.

Multiple Concurrent Requests with `forkJoin()` in Angular

The code sample in the previous section reads the contents of a single file. However, in some cases you might need to load data from multiple sources, and delay the post-loading logic until all the data has loaded. `Observables` provide a method called `forkJoin()` to "wrap" multiple `Observables`, and make multiple concurrent `http.get()` requests. Note that the operation fails if any individual request fails. The `subscribe()` method sets the handlers on the entire set of `Observables`.

Copy the directory `ForkJoin` from the companion disc into a convenient location. Listing 4.8 displays the contents of `app.component.ts` that illustrates how to reference the custom component `FileService`, which reads the contents of `employees.json` and `customers.json`.

LISTING 4.8: app.component.ts

```
import {Component}        from '@angular/core';
import {FileService}      from './file.service';

@Component({
  selector: 'app-root',
  template:'
  <h2>Angular2 HTTP and Observables</h2>
  <h3>Some of our Employees</h3>
  <ul>
    <li *ngFor="let emp of employees">
      {{emp.fname}} {{emp.lname}} lives in {{emp.city}}
```

```
    </li>
  </ul>

  <h3>Some of our Customers</h3>
  <ul>
    <li *ngFor="let cust of customers">
      {{cust.fname}} {{cust.lname}} lives in {{cust.city}}
    </li>
  </ul>
  `
})
export class AppComponent {
  public employees;
  public customers;

  constructor(private _fileService: FileService) { }

  ngOnInit() {
    this.getBothFiles();
  }

  getBothFiles() {
    this._fileService.getBothFiles().subscribe(
      data => {
        this.customers = data[0]
        this.employees = data[1]
      }
      // error/completion callbacks are optional, and console
      // messages appear if the Observable is in an error state
    );
  }
}
```

The `template` property in Listing 4.8 contains an unordered list of employees and an unordered list of customers, both of which are retrieved via the method `getBothFiles()` in the custom `AppComponent` component. Notice that the retrieved data is an array of two elements, where the first element contains customer-related data and the second element contains employee-related data.

The `getBothFiles()` method is invoked in the `ngOnInit()` lifecycle method, and the actual `HTTP GET` request is performed in a method (also called `getBothFiles()`) that is defined in the `FileService` custom component.

Listing 4.9 displays the contents of `file.service.ts` that defines the class `FileService`. This class contains a method `getBothFiles()` that uses `forkJoin()` to read data from two `JSON`-based files.

LISTING 4.9: file.service.ts

```
import {Injectable}       from '@angular/core';
import {Http, Response} from '@angular/http';
import {Observable}       from 'rxjs/Rx';

@Injectable()
export class FileService {
  constructor(private http:Http) { }

  // http.get() loads one JSON file
  getEmployees() {
    return this.http.get('/src/app/employees.json')
               .map((res:Response) => res.json());
  }

  getBothFiles() {
    return Observable.forkJoin(
      this.http.get('/src/app/customers.json')
          .map((res:Response) => res.json()),
      this.http.get('/src/app/employees.json')
          .map((res:Response) => res.json())
    );
  }
}
```

Listing 4.9 contains the `getBothFiles()` method, which invokes the `forkJoin()` method of the `Observable` class to retrieve the `JSON` data in the file `src/app/customers.json` and the file `src/app/employees.json`.

The following code snippet appears three times in Listing 4.9, and in every case it returns `JSON`-formatted data inside an `Observable`:

```
.map((res:Response) => res.json())
```

The employee-related data is displayed in Listing 4.6, and Listing 4.10 displays the contents of `customers.json` that contains customer-related information.

LISTING 4.10: customers.json

```
[
{"fname":"Paolo","lname":"Friulano","city":"Maniago"},
{"fname":"Luigi","lname":"Napoli","city":"Vicenza"},
{"fname":"Miko","lname":"Tanaka","city":"Yokohama"},
```

```
{"fname":"Yumi","lname":"Fujimoto","city":"Tokyo"}
]
```

Launch the Angular application via the `ng serve` command. You will see the following output in a browser session at the address `localhost:4200`:

Angular2 HTTP and Observables

Some of our Employees

- Paolo Friulano lives in Maniago
- Luigi Napoli lives in Vicenza
- Miko Tanaka lives in Yokohama
- Yumi Fujimoto lives in Tokyo

Some of our Customers

- Jane Jones lives in San Francisco
- John Smith lives in New York
- Dave Stone lives in Seattle
- Sara Edson lives in Chicago

Now that you have a basic grasp of how to work with `Observables` in Angular applications, let's see how to represent `JSON` data in a TypeScript interface.

TypeScript Interfaces in Angular

The `JSON` file `customers.json` contains record-like information about customers, each of which can be modeled via something called an interface, which exists in many programming languages. Interfaces are somewhat analogous to a `struct` in languages such as Apple Swift, Google Go, or the C programming language.

In general, a TypeScript interface specifies a list of variables and their type, along with one or more methods. A TypeScript interface is a useful construct whenever you want to associate logically related data items (i.e., pieces of information). Many examples of entities with multiple data items are available, such as customers, employees, students, purchase orders, and so forth. You can model all of them (and many others) with a TypeScript interface, an example of which is discussed in the next section.

A Simple TypeScript Interface

Listing 4.11 displays the contents of `employee.ts` that contains an interface for a hypothetical employee. A real-life example would contain many other pieces of information (and methods) in such an interface, whereas this highly simplified example is intended to illustrate how to use TypeScript interfaces in Angular.

LISTING 4.11: employee.ts

```
export interface Employee {
  fname: string,
  lname: string,
  city:  string
}
```

Notice that the structure of the interface `Employee` in Listing 4.11 matches the contents of the rows in `Employees.json`.

JSON Data and TypeScript Interfaces

Listing 4.12 displays the contents of `employees.ts` that contains an array of `Employee` instances. The file `employees.ts` illustrates how to populate an array with hard-coded data values that conform to a TypeScript interface. Note that Listing 4.10 is not used in the current code sample. Its purpose is to merely demonstrate that it's possible to load "seed data" (which can be useful for testing purposes) from a static file.

LISTING 4.12: employees.ts

```
import {Employee} from './employee';

export const EMPLOYEES : Employee[] = [
  {"fname":"Jane","lname":"Jones","city":"San Francisco"},
  {"fname":"John","lname":"Smith","city":"New York"},
  {"fname":"Dave","lname":"Stone","city":"Seattle"},
  {"fname":"Sara","lname":"Edson","city":"Chicago"}
];
```

An Angular Application with a TypeScript Interface

This section discusses how to create an Angular application that uses a TypeScript interface.

Create an Angular application called `ReadJSONFileTS` by cloning the Angular application `ReadJSONFile` and then make two small changes to the contents of `app.component.ts` in `ReadJSONFileTS`.

The first change is to include the following `import` statement:

```
import {Employee} from './employee';
```

The second change is to replace the code snippet in bold in Listing 4.10 with the following code snippet, which declares the employees variable to be an array of objects that conform to the `Employee` interface:

```
employees : Employee[];
```

Launch the application and you will see the same results. Keep in mind that the only significant difference in this project involves a TypeScript interface. Although the use of a TypeScript interface in this example does not provide a significant advantage, consider the case of Angular applications that contain multiple TypeScript interfaces, some of which might contain dozens of elements. In such scenarios, TypeScript interfaces simplify the task of keeping track of related data, which is useful when you need to perform create, read, update, and delete (CRUD) operations.

There are some useful facts to keep in mind about TypeScript interfaces. First, they can contain method definitions that are implemented in a class. Second, a TypeScript interface can extend an existing interface (so it's possible to create a hierarchical structure). Third, a TypeScript interface can extend a TypeScript class. Although this chapter does not contain examples that illustrate any of these additional features, you can perform an online search to find relevant code samples and tutorials.

Getting GitHub User Data in Angular

Copy the directory `GithubUsers` from the companion disc into a convenient location. Listing 4.13 displays the contents of `app.component.ts` that illustrates how to make an `HTTP GET` request to retrieve information about GitHub users.

LISTING 4.13: app.component.ts

```
import { Component }     from '@angular/core';
import {Inject}          from '@angular/core';
```

```
import {Http}            from '@angular/http';
import {HTTP_BINDINGS}   from '@angular/http';
import 'rxjs/add/operator/map';

@Component({
   selector: 'app-root',
   template: '<button (click)="httpRequest()">User Info
                                                    </button>'
})
export class AppComponent {
  githubData : any;

  constructor(@Inject(Http) public http:Http) {}

  httpRequest() {
        this.http.get('https://api.github.com/users/
                                              ocampesato')
      .map(res => res.json())
      .subscribe(
        data => this.githubData = data,
        err => console.log('error'),
        () => this.userInfo()
      );
  }

  userInfo() {
     console.log("followers  = "+this.githubData.followers);
     console.log("following  = "+this.githubData.following);
     console.log("created_at = "+this.githubData.created_at);
  }
}
```

Listing 4.13 contains code that is similar to the code in Listing 4.5. The difference is in the `httpRequest()` method, which makes an `HTTP GET` request from a live endpoint instead of reading data from a file. Another difference is that the only data that is displayed is shown in the console tab of the browser.

Launch the code in this section; in the console tab, you will see something similar to the following output:

```
followers  = 16
following  = 2
created_at = 2011-07-14T23:06:31Z
```

The code in this section gives you a starting point for displaying additional details regarding a user, and displays the information in a more pleasing manner.

HTTP GET Requests with a Simple Server

This section shows you how to work with the command line utility `json-server`, which can serve JSON-based data. This program performs the function of a very simple server: clients can make `GET` requests to retrieve `JSON` data from a server. Moreover, a simple command in the console where `json-server` was launched enables you to save the in-memory data to a file.

Although `json-server` does not perform the functions of a node-based application that contains Express and MongoDB, `json-server` is a convenient program that helps you learn how an Angular application can interact with a file server.

You need to perform the following steps:

- Step 1: Install `json-server`.
- Step 2: Launch `json-server`.
- Step 3: Launch the Angular application.

Install `json-server` via the following command:

```
[sudo] npm install -g json-server
```

Navigate to the directory that contains the `JSON` file `posts.json` and invoke this command:

```
json-server posts.json
```

The preceding command launches a file server at port `3000` and reads the contents of `posts.json` into memory, making that data available to `HTTP GET` requests.

Copy the directory `JSONServerGET` from the companion disc into a convenient location. Listing 4.14 displays the contents of `app.component.ts` that illustrates how to make an `HTTP GET` request to retrieve data from a file server.

LISTING 4.14: app.component.ts

```
import {Component}       from '@angular/core';
import {Inject}          from '@angular/core';
import {Http}            from '@angular/http';
import {HTTP_BINDINGS}   from '@angular/http';
import 'rxjs/add/operator/map';
```

```
@Component({
  selector: 'app-root',
  template: `
    <button (click)="httpRequest()">Get Information</button>
    <div>
      <li *ngFor="let post of postData">
        {{post.author}}
        {{post.title}}
      </li>
    </div>
  `
})
export class AppComponent {
  postData = "";

  constructor(@Inject(Http) public http:Http) {
  }

  httpRequest() {
    this.http.get('http://localhost:3000/posts')
      .map(res => res.json())
      .subscribe(
        data => this.postData = JSON.stringify(data),
        err => console.log('error'),
        () => this.postInfo()
      );
  }

  postInfo() {
     //------------------------------------------
     // the 'eval' statement is required in order to
     // convert the data retrieved from json-server
     // to an array of JSON objects (else an error)
     //------------------------------------------
     var myObject = eval('(' + this.postData + ')');
     console.log("myObject = "+JSON.stringify(myObject));
     this.postData = myObject;
  }
}
```

Listing 4.14 contains code that is similar to earlier code samples. The first difference involves the details of the unordered list that is displayed in the template property.

The second difference involves the local endpoint `http://localhost:3000/users` in the `HTTP GET` request. This endpoint provides `JSON` data via the `json-server` that is listening on port `3000`.

Listing 4.15 displays the contents of `posts.json` that is retrieved during the `HTTP GET` request in Listing 4.14.

LISTING 4.15: posts.json

```
{
  "posts": [
    {"id": 100,"title": "json-server","author": "smartguy"},
    {"id": 200,"title": "pizza-maker","author": "chicago"},
    {"id": 300,"title": "good-beer",  "author": "escondido"}
  ]
}
```

The next section shows you how to make an `HTTP POST` request to a local file server in an Angular application.

HTTP POST Requests with a Simple Server

This section shows you how to make an `HTTP POST` request with the utility `json-server`, which can serve `JSON`-based data. Keep in mind that the server in this code sample only handles basic data requests; "universal" JavaScript (sometimes called "isomorphic" JavaScript) is not covered in this chapter.

Navigate to the `src/app` subdirectory, which contains the `JSON` file `authors.json` and launch this command:

```
json-server authors.json
```

The preceding command launches a file server at port `3000` and reads the contents of `authors.json` into memory, making that data available to `HTTP GET` requests.

Now copy the directory `JSONServerPOST` from the companion disc into a convenient location. Listing 4.16 displays the contents of `app.component.ts` that illustrates how to make an `HTTP POST` request to a local file server.

LISTING 4.16: app.component.ts

```
import { Component }    from '@angular/core';
import {Inject}         from '@angular/core';
import {Http}           from '@angular/http';
import 'rxjs/add/operator/map';
declare var $: any;

@Component({
   selector: 'app-root',
```

```
    template: `
      <button (click)="getEmpData()">Author Info</button>
      <div>
        <table>
          <thead *ngIf="foundData">
            <th>AUTHORID</th>
            <th>Title</th>
            <th>Author</th>
          </thead>
          <tbody>
            <tr *ngFor="let author of authorData">
              <td>{{author.id}}</td>
              <td>{{author.title}}</td>
              <td>{{author.author}}</td>
            </tr>
          </tbody>
        </table>
        <button (click)="postAuthorData()">Add Author</button>
      </div>
    `
})
export class AppComponent {
  foundData   = false;
  authorData  = [];
  currData    = {};
  idIncr      = 100;
  newAuthorId = 0;
  newTitle    = "";
  newAuthor   = "";
  largestId   = 0;

  constructor(@Inject(Http) public http:Http) {}

  postAuthorData() {
    this.newAuthorId = 0+this.largestId+this.idIncr;
    this.newTitle    = "The Book of "+this.newAuthorId;
    this.newAuthor   = "My New Title"+this.newAuthorId;

    var postNewAuthor = {id:this.newAuthorId,
                         title:this.newTitle,
                         author:this.newAuthor};

//console.log("postNewAuthor: "+JSON.stringify(postNewAuthor));

    $.post("http://localhost:3000/authors",
       postNewAuthor,
       function(result, textStatus, jqXHR) {
  //console.log("2returned result: "+JSON.stringify(result));
           this.authorData.push(postNewAuthor);
       }.bind(this),"json")
```

```
        .fail(function(jqXHR, textStatus, errorThrown) {
  console.log("error: "+errorThrown+" textStatus:
                                                   "+textStatus);
      });
  }

  getAuthorData() {
    this.http.get('http://localhost:3000/authors')
      .map(res => res.json())
      .subscribe(
        data => this.authorData = data,
        err => console.log('error'),
        () => this.authorInfo()
      );
  }

  authorInfo() {
     this.largestId =
         parseInt(this.authorData[this.authorData.length-1].
                                                     id,10);

   //console.log("largestId    = "+ this.largestId);
   //console.log("authorData1 = "+ JSON.stringify(this.
                                                authorData));

     this.foundData = true;
  }
}
```

Listing 4.16 contains the same `import` statements as Listing 4.14, followed by a `template` property that displays a table of author-based data. When users click the `<button>` element, the `postAuthorData()` adds a hard-coded new author to the list of authors. This method performs a standard jQuery `POST` request instead of using an `Observable`. Note that this method increments the value of the `id` property of each author so that they are treated as distinct authors (even though the names of the new users are almost the same).

On the other hand, the `getAuthorData()` method does involve an `Observable` for retrieving author-related data (shown in Listing 4.17) from the file server that is running on port `3000`.

NOTE *The current functionality only supports the insertion of one new author. As an exercise, modify the code to support the insertion of multiple new authors.*

One other minor point: The browser is reloaded after each invocation of the `postAuthorData()` method, so you need to click the "Author Info" button to see the newly added author.

Listing 4.17 displays a portion of the contents of `authors.json`, whose contents are displayed in this Angular application.

LISTING 4.17: authors.json

```
{
  "authors": [
    {
      "id": 100,
      "title": "json-server",
      "author": "typicode"
    },
    {
      "id": 200,
      "title": "pizza-maker",
      "author": "chicago"
    },
// sections omitted for brevity
    {
      "id": "900",
      "title": "The Book of 900",
      "author": "My New Title900"
    }
  ]
}
```

As you can see, Listing 4.17 is a very simple collection of JSON-based data items, where each item contains the elements `id`, `title`, and `author`.

This concludes the portion of the chapter involving code samples that use `json-server` as a local file server. The next section of this chapter discusses how routing is supported in Angular.

Routing in Angular

Web applications can have different sections that correspond to different URLs, and supporting those sections programmatically is called *routing*. When you see a Web page that contains tabs or a set of horizontal links that display different sections of an application, it's quite likely that the Web page is using some form of routing.

For instance, a Web page might provide an "about" section, a "login" section, and an "orders" section. Suppose that you want to allow access to the orders section only after users have logged into the application. Routes can restrict access to the orders section, and maintain state in order to simplify the code that makes the transitions between sections.

Routing in Angular applications involves adding a “mapping” between routes and actions in `app.module.ts`, an example of which is shown in the variable `appRoutes` below:

```
const appRoutes: Routes = [
  {path: 'about',   component: AboutComponent},
  {path: 'login',   component: LoginComponent},
  {path: 'contact', component: ContactComponent}
]

@NgModule({
  imports:      [ BrowserModule,
                  FormsModule,
                  RouterModule.forRoot(appRoutes),
                ],
  declarations: [ AppComponent ],
  bootstrap:    [ AppComponent ]
})
```

The preceding code block defines the variable `appRoutes`, which specifies three routes—`about`, `login`, and `contact`—which are mapped to the components `AboutComponent`, `LoginComponent`, and `ContactComponent`, respectively.

Regardless of the link that users click, the relevant component is displayed in the `<router-outlet>` element (specified in the template definition in `app.component.ts`). When users click a link, all existing content in `<router-outlet>` is removed and replaced with the component that is associated with the currently clicked link.

To use a tab-based analogy, when users click a tab, the current screen contents are replaced by the contents of the most recently clicked tab.

A more generalized scenario of configuring a set of routes supports rerouting and specifying parameters, as shown here:

```
[
  { path: 'home',          component: HomeComponent },
  { path: 'courses',       component: CourseListComponent },
  { path: 'course/:id',    component: CourseDetailsComponent },
  { path: '', redirectTo: 'home', pathMatch: 'full' },
  { path: '**',            component: UnknownPageComponent }
]
```

More detailed information about Angular routes is located here:

https://angular.io/docs/ts/latest/guide/router.html

Keep in mind that routing-related code has changed from earlier versions of Angular, and might also change in the future, so it's better to check more recent posts on the Stack Overflow site as well as the online documentation.

The next section contains a simple yet complete code sample that uses Angular routes.

A Routing Example in Angular

The code sample in this section shows you how to set up basic routing in Angular applications. Copy the directory `BasicRouting` from the companion disc to a convenient location. Notice the following code snippet above the `<body>` element in `index.html`:

```
<base href="/">
```

The preceding code snippet is required for any Angular application that involves routing-related functionality; in fact, Angular relies on this tag to determine how to construct its routing information.

Listing 4.18 displays the contents of `app.component.ts` that illustrates a simple example of routing in Angular.

LISTING 4.18: app.component.ts

```
import {Component}          from '@angular/core';
import {ROUTER_PROVIDERS}   from '@angular/router';
import {ROUTER_DIRECTIVES}  from '@angular/router';
import {RouteConfig}        from '@angular/router';

import {About} from './about';
import {Login} from './login';
import {Users} from './users';

@Component({
  selector: 'app-root',
  template: `
    <h1 class="title">Angular Router</h1>
    <nav>
      <a [routerLink]="['About']">About</a>
      <a [routerLink]="['Login']">Login</a>
```

```
      <a [routerLink]="['Users']">Users</a>
    </nav>
    <router-outlet></router-outlet>
  `
})
@RouteConfig([
  {path: '/about', name: 'About', component: About},
  {path: '/login', name: 'Login', component: Login,
                                             useAsDefault:true},
  {path: '/users', name: 'Users', component: Users}
])
export class AppComponent { }
```

Listing 4.18 contains three `import` statements for route-related functionality, followed by `import` statements to access the custom components `About`, `Login`, and `Users`.

The `template` property in Listing 4.18 contains a `<nav>` element that comprises three anchor elements to set up the links to the three custom components. The template property also contains the `<router-outlet>` element, which is where the output for each custom component will be rendered.

The last portion of Listing 4.17 contains the `RouteConfig` mapping that makes the association between the custom components and the path elements.

Listing 4.19 displays the contents of `about.ts` that is a component for allowing users to log into the application.

LISTING 4.19: about.ts

```
import {Component} from '@angular/core';

@Component({
   selector: 'app-root',
   template: `
     <div>
       <p>This 'About' page is part of the Routing example.
       </p>
     </div>
   `
})
export class About { }
```

Listing 4.19 is straightforward: The `template` property contains a `<p>` element with a text string that is displayed in the `<router-outlet>` element of the root component.

Listing 4.20 displays the contents of `login.ts` that is a component for allowing users to log in to the application.

LISTING 4.20: login.ts

```
import {Component} from '@angular/core';
import {User}      from './user';

@Component({
   selector: 'app-root',
   template: `
     <div>
       <br />
       <label for="uname">Username:</label>
       <input #uname> <br />
       <label for="passwd">Password:</label>
       <input #passwd> <br />
       <button (click)="clickMe(uname.value,passwd.value)">
         Login
       </button>
     </div>
   `
})
export class Login {
   clickMe(name,pwd) {
      // insert your code to validate the username/password
      console.log("Perform validation logic: "+name+" "+pwd);
   }
}
```

Listing 4.21 contains a `template` property that simulates the login process for a user. Input fields for the user name and password are supplied, and when users click the `<button>` element, the `clickMe()` method is invoked but no actual validation is performed in that method.

LISTING 4.21: users.ts

```
import {Component} from '@angular/core';
import {User}      from './user';

@Component({
   selector: 'app-root',
   template: `
     <div>
       <ul>
         <li *ngFor="let user of users"
                                     (click)="onSelect(user)">
```

```
            {{user.fname}}
          </li>
        </ul>
      </div>
    `
})
export class Users {
  users = [
            new User("Jane"),
            new User("Dave"),
            new User("Tom")
           ];

   onSelect(user) {
      console.log("Selected user: "+JSON.stringify(user));
      var index = this.users.indexOf(user);
      this.users.splice(index,1);
   }
}
```

Listing 4.21 contains a `template` property that displays the list of names of the users in the `users` array, which is initialized in the `Users` class. Each user is created as an instance of the `User` class, which currently contains only the first name of a user; a realistic example would obviously contain many more user-related properties.

Listing 4.22 displays the contents of `user.ts` that is a component for allowing users to register themselves in the application.

LISTING 4.22: user.ts

```
import {Component} from '@angular/core';

@Component({
   selector: 'user',
   template: '',
})
export class User {
   fname:string;

   constructor(fname:string) {
      this.fname = fname;
   }
}
```

Listing 4.22 defines the `User` class, which contains a constructor for initializing the first name of a user.

Angular Routing with Webpack

If you use Webpack in conjunction with Angular applications, you can perform Angular routing with Webpack using the following router loader:

https://www.npmjs.com/package/angular2-router-loader

The GitHub repository is located here:

https://github.com/brandonroberts/angular2-router-loader

An article that describes how to use the preceding router loader is located here:

https://medium.com/@daviddentoom/angular-2-lazy-loading-with-webpack-d25fe71c29c1#.8srgkl44h

Summary

This chapter started by showing you how to make an `HTTP GET` request to read the contents of a `JSON`-based file. Next you learned how to make an `HTTP GET` request to retrieve information about a `Github` user, and how to make `HTTP POST` requests. You also learned how to retrieve `JSON` data from a website, convert that data to a `Promise`, and then convert the `Promise` into an `Observable`. Finally, you learned how to set up routing in an Angular application.

CHAPTER 5

FORMS, PIPES, AND SERVICES

This chapter shows you how to create Angular applications that use Angular `Forms`, `Pipes`, and Services. The code samples rely on an understanding of functionality that is discussed in earlier chapters, such as how to make HTTP requests in Angular.

The first section in this chapter contains Angular applications that use Angular `Controls` and `Control Groups`. This section also provides an example of an Angular application that contains a form that makes `HTTP GET` requests, which enhances the example involving `HTTP`-related functionality in Chapter 4.

The second part of this chapter discusses Angular `Pipes` (the counterpart to `Filters` in Angular 1.x). You will also learn about `async` pipes in this section, which can eliminate the need for defining instance variables and also reduce the likelihood of memory leaks in Angular applications.

Overview of Angular Forms

An Angular `FormControl` represents a single input field, a `FormGroup` consists of multiple logically related fields, and an `NgForm` component represents a `<form>` element in an HTML Web page. The `ngSubmit` action for submitting a form has this syntax:

```
(ngSubmit)="onSubmit(myForm.value)".
```

Note that `NgForm` provides the `ngSubmit` event, whereas you must define the `onSubmit()` method in the component class. The expression `myForm.value` consists of the key/value pairs in the form. Later in the chapter you will see examples involving these controls, as well as `FormBuilder`, which supports additional useful functionality.

Angular also supports template-driven forms (with a `FormsModule`) and reactive forms (with a `ReactiveFormsModule`), both of which belong to `@angular/forms`. However, Reactive Forms are synchronous whereas template-driven forms are asynchronous.

Reactive forms

Reactive forms involve explicit management of the data flowing between a non-user interface (UI) data model and a UI-oriented form model that retains the states and values of the HTML controls on screen. Reactive forms offer the ease of using reactive patterns, testing, and validation.

Reactive Forms involve the creation of a tree of Angular form control objects in the component class `app.component.ts`, which are also bound to native form control elements in the component template `app.component.html`.

The component class has access to the data model and the form control structure, which enables you to propagate data model values into the form controls and retrieve user-supplied values in the HTML controls. The component can observe changes in form control state and react to those changes. One advantage of working with form control objects directly is that value and validity updates are always synchronous and under your control. You won't encounter the timing issues that sometimes plague a template-driven form and reactive forms can be easier to unit test. Since reactive forms are created directly via code, they are always available, which enables you to immediately update values and "drill down" to descendant elements.

Template-Driven Forms

Template-driven forms involve placing HTML form controls (such as `<input>`, `<select>`, and so forth) in the component template. In addition, the form controls are bound to data model properties in the component via directives such as `ngModel`.

Note that Angular directives create Angular form objects based on the information in the provided data bindings. Angular uses `ngModel` to handle the transfer of data values, and also updates the mutable data model with user changes as they happen. Consequently, the `ngModel` directive does not belong to the `ReactiveFormsModule`.

Before delving into the material in this section, the companion disc contains the Angular application `MasterForm`, which has form-related code. Although this code sample does not use Angular FormGroups, you might find some useful features in the code.

The next section shows you how to use the Angular `ngForm` component to create a form "the Angular way." Then you will see an example that shows you how to use an Angular `FormGroup` in an Angular Application.

An Angular Form Example

Copy the directory `NGForm` from the companion disc into a convenient location. Listing 5.1 displays the contents of `app.component.ts` that illustrates how to use `<input>` elements with an `ngModel` attribute in an Angular application.

LISTING 5.1: app.component.ts

```
import { Component } from '@angular/core';

@Component({
  selector: 'app-root',
  template: `
    <div>
      <h2>A Sample Form</h2>
      <form #f="ngForm"
            (ngSubmit)="onSubmit(f.value)"
            class="ui form">
        <div class="field">
          <label for="fname">fname</label>
          <input type="text"
                 id="fname"
                 placeholder="fname"
                 name="fname" ngModel>

          <label for="lname">lname</label>
          <input type="text"
                 id="lname"
                 placeholder="lname"
                 name="lname" ngModel>
        </div>

        <button type="submit">Submit</button>
      </form>
    </div>
  `
})
```

```
export class AppComponent {
  myForm: any;

  onSubmit(form: any): void {
    console.log('you submitted value:', form);
  }
}
```

Listing 5.1 defines a template property that contains a `<form>` element with two `<div>` elements, each of which contains an `<input>` element. The first `<input>` element is for the first name and the second `<input>` element is for the last name of a new user.

Angular provides the `NgModel` directive, which enables you to use the instance variable `myForm` in an Angular form. For example, the following code snippet specifies `myForm` as the control group for the given form:

```
<form [ngModel]="myForm"
  (ngSubmit)="onSubmit(myForm.value)"
```

Notice that `onSubmit` specifies `myForm` and that a `Control` is bound to the input element.

NOTE *Add the attribute `novalidate` to the `<form>` element to disable browser validation.*

Listing 5.2 displays the contents of `app.module.ts` that imports a `FormsModule` and includes it in the `imports` property.

LISTING 5.2: app.module.ts

```
import { NgModule }       from '@angular/core';
import { FormsModule }    from '@angular/forms';
import { BrowserModule } from '@angular/platform-browser';
import { AppComponent }   from './app.component';

@NgModule({
  imports:      [ BrowserModule, FormsModule ],
  declarations: [ AppComponent ],
  bootstrap:    [ AppComponent ]
})
export class AppModule { }
```

Listing 5.2 is straightforward: It contains two lines (shown in bold) involving the `FormsModule` that are required for this code sample.

Data Binding and ngModel

Angular supports three types of binding in a form: no binding, one-way binding, and two-way binding. Here are some examples:

```
<!-- no binding -->
<input name="fname" ngModel>

<!-- one-way binding -->
<input name="fname" [ngModel]="fname">

<!-- two-way binding -->
<input name="fname" [ngModel]="fname"
       (ngModelChange)="fname=$event">

<!-- two-way binding -->
<input name="fname" [(ngModel)]="fname">
```

The one-way binding example will look for the `fname` property in the associated component and initialize the `<input>` field with the value of the `fname` property.

The two-way binding example fires the `ngModelChange` event when users alter the value of the `<input>` field, which causes an update to the `fname` property in the component, thereby ensuring that the input value and its associated component value are the same. You can also replace the value of `ngModelChange` with the output of a function (e.g., capitalizing the text string that users enter in the input field).

The second example of two-way data binding uses the "banana in a box" syntax, which is a shorthand way of achieving the same result as the first two-way data binding example. However, this syntax does not support the use of a function that is possible with the longer syntax for two-way data binding.

Third-Party UI Components

Valor provides Bootstrap components for Angular, and its home page is located here:

https://valor-software.com/ng2-bootstrap/#/

This module provides Twitter Bootstrap components and a date picker, time picker, rating, and typeahead (among other components), and also works with Bootstrap version 3 and version 4.

Alternatively, you can use `ng2-bootstrap` in Angular applications. There are three steps required to install and use `ng2-bootstrap` in an Angular application. First, install `ng2-bootstrap` with this command:

```
npm install ng2-bootstrap --save
```

Second, add this snippet to `index.html` to reference Bootstrap styles:

```
<link href="https://maxcdn.bootstrapcdn.com/bootstrap/3.3.7/
css/bootstrap.min.css" rel="stylesheet">
```

Third, insert a new entry into the `styles` array in `angular-cli.json`:

```
"styles": [
   "../node_modules/bootstrap/dist/css/bootstrap.min.css",
   "styles.css",
],
```

For additional instructions and code samples, navigate to this URL:

https://github.com/valor-software/ng2-bootstrap

The next portion of this chapter shows you how to work with forms in "the Angular way."

Angular Forms with FormBuilder

The `FormBuilder` class and the `FormGroup` class are built-in Angular classes for creating forms. `FormBuilder` supports the `control()` function for creating a `FormControl` and the `group()` function for creating a `FormGroup`.

Copy the directory `FormBuilder` from the companion disc to a convenient location. Listing 5.3 displays the contents of `app.component.ts` that illustrates how to use an Angular form in an Angular application.

LISTING 5.3: app.component.ts

```
import { Component }   from '@angular/core';
import { FormBuilder } from '@angular/forms';
import { FormGroup }   from '@angular/forms';

@Component({
  selector: 'app-root',
  template: `
```

```
    <div>
      <h2>A FormBuilder Form</h2>

      <form [formGroup]="myForm"
            (ngSubmit)="onSubmit(myForm.value)"
            class="ui form">

        <div class="field">
          <label for="fname">fname</label>
          <input type="text"
                 id="fname"
                 placeholder="fname"
                 [formControl]="myForm.controls['fname']">
        </div>

        <div class="field">
          <label for="lname">lname</label>
          <input type="text"
                 id="lname"
                 placeholder="lname"
                 [formControl]="myForm.controls['lname']">
        </div>

        <button type="submit">Submit</button>
      </form>
    </div>
    `
})
export class AppComponent {
//myForm: any;
  myForm: FormGroup;

  constructor(fb: FormBuilder) {
    this.myForm = fb.group({
      'fname': ['John'],
      'lname': ['Smith']
    });
  }

  onSubmit(value: string): void {
    console.log('you submitted value:', value);
  }
}
```

Listing 5.3 contains a <form> element with two <div> elements, each of which contains an <input> element. The first <input> element is for the first name and the second <input> element is for the last name of a new user.

In Listing 5.3, `FormBuilder` is injected into the constructor, which creates an instance of `FormBuilder` that is assigned to the `fb` variable in the constructor. Next, `myForm` is initialized by invoking the `group()` method that takes an object of key/value pairs. In this case, `fname` and `lname` are keys, and both of them appear as `<input>` elements in the `template` property. The values of these keys are optional initial values.

Obviously you can add many other properties inside the `group()` method (such as address-related fields). Moreover, you can add a different form for each new entity. For example, you could create separate forms for a `Customer`, `PurchaseOrder`, and `LineItems`.

Angular Reactive Forms

Copy the directory `ReactiveForm` from the companion disc to a convenient location. Listing 5.4 displays the contents of `app.component.ts` that illustrates how to define a reactive Angular form in an Angular application.

LISTING 5.4: app.component.ts

```
import { Component }   from '@angular/core';
import { FormBuilder } from '@angular/forms';
import { FormGroup }   from '@angular/forms';
import { FormControl } from '@angular/forms';

@Component({
  selector:    'app-root',
  templateUrl: './app.component.html',
  styleUrls:   ['./app.component.css']
})
export class AppComponent {
   userForm: FormGroup;
   disabled:boolean;

   constructor(fb: FormBuilder) {
     this.userForm = fb.group({
        name:     'Jane',
        email:    'jsmith@yahoo.com',
        address: fb.group({
          city:  'San Francisco',
          state: 'California'
        })
     });
   }
```

```
    onFormSubmitted(theForm : FormGroup) {
       console.log("name  = "+theForm.controls['name'].value);
       console.log("email = "+theForm.controls['email'].value);
       console.log("city  = "+theForm.get('address.city').
                                                     value);
       console.log("city  = "+theForm.get('address.state').
                                                     value);
    }
}
```

Listing 5.4 contains the usual `import` statements. Notice how the variable `userForm`, which has type `FormBuilder`, is initialized in the constructor. In addition to two text fields, `userForm` contains the `address` element, which also has type `FormBuilder`.

Listing 5.5 displays the contents of `app.module.html` with an Angular form that contains `<input>` elements that correspond to the fields in the `userForm` variable.

LISTING 5.5: app.component.html

```
<form [formGroup]="userForm"
                    (ngSubmit)="onFormSubmitted(userForm)">
  <label>
    <span>Name</span>
    <input type="text" formControlName="name"
placeholder="Name" required>
  </label>

  <div>
    <label>
      <span>Email</span>
      <input type="email" formControlName="email"
placeholder="Email" required>
    </label>
  </div>

  <div formGroupName="address">
    <div>
      <label>
        <span>City</span>
        <input type="text" formControlName="city"
                                placeholder="City" required>
      </label>
    </div>
    <label>
      <span>Country</span>
      <input type="text" formControlName="state"
                                placeholder="State" required>
```

```
      </label>
    </div>
      <br />
    <input type="submit" [disabled]="userForm.invalid">
</form>
```

Listing 5.5 contains very simple HTML markup that enables users to change the default values for each of the input fields.

Listing 5.6 displays the updated contents (shown in bold) of `app.module.ts`.

LISTING 5.6: app.module.ts

```
import { BrowserModule } from '@angular/platform-browser';
import { NgModule }             from '@angular/core';
import { FormsModule }          from '@angular/forms';
import { HttpModule }           from '@angular/http';
import { ReactiveFormsModule } from '@angular/forms';
import { AppComponent }         from './app.component';

@NgModule({
  declarations: [
    AppComponent
  ],
  imports: [
    BrowserModule,
    FormsModule,
    HttpModule,
    ReactiveFormsModule
  ],
  providers: [],
  bootstrap: [AppComponent]
})
export class AppModule { }
```

Listing 5.6 contains one new line of code: an `import` statement for `ReactiveFormsModule` (which can be combined with the `import` statement for `FormsModule`), which is also referenced in the `imports` property.

FormGroup versus FormArray

As you now know, a `FormGroup` aggregates the values of `FormControl` elements into one object, where the control name is the key. Angular also supports `FormArray` (a "variation" of `FormGroup`), which aggregates the values of `FormControl` elements into an array.

FormGroup data is serialized as an array, whereas FormArray data is serialized as an object). If you do not know how many controls are in a given group, consider using a FormArray (otherwise use a FormGroup). The following link contains an example of using a FormArray:

https://alligator.io/angular/reactive-forms-formarray-dynamic-fields/

Other Form Features in Angular

The preceding section gave you a glimpse into the modularized style of Angular forms, and this brief section highlights some additional form-related features in Angular, such as the following:

- Form validation
- Custom validators
- Nested forms
- Dynamic forms
- Template-driven forms

Validators enable you to perform validation on form fields, such as specifying mandatory fields and the minimum and maximum lengths of fields. You can also specify a regular expression that a field must match, which is very useful for zip codes, email addresses, and so forth. Alternatively, you can specify validators programmatically.

Angular forms also provide event listeners that detect various events pertaining to the state of a form, as shown in the following code snippets:

```
{{myform.form.touched}}
{{myform.form.untouched}}
{{myform.form.pristine}}
{{myform.form.dirty}}
{{myform.form.valid}}
{{myform.form.invalid}}
```

For example, the following <span> element is displayed if one or more form fields is invalid:

```
<span *ngIf="!myform.form.valid">The Form is Invalid</span>
```

You can also display error messages using the *ngIf directive to display the status of a specific field, as shown here:

```
<label>
  <span>First Name</span>
```

```
<input type="text" formControlName="fname"
                                    placeholder="First Name">
  <p *ngIf="userForm.controls.fname.errors">
    This value is invalid
  </p>
</label>
```

You can find an example of a dynamic Angular form is here:

https://angular.io/docs/ts/latest/cookbook/dynamic-form.html

You can find an example of a template-driven Angular form here:

https://toddmotto.com/angular-2-forms-template-driven

Instead of using plain Cascading Style Sheets (CSS) for styling effects for field-related error messages, consider using something like ng2-bootstrap or Bootstrap (discussed briefly in the next section).

Angular Forms and Bootstrap 4 (optional)

Although Bootstrap is external to Angular, you probably want to add some styling effects to your Angular applications, especially if they contain an Angular form with many controls. For instance, the `ReactiveForm` application in a previous section provides a very plain-looking UI, and Bootstrap can help you greatly improve its appearance.

However, you are not bound to Bootstrap, so this section only shows you some visual effects that are very easy to create with Bootstrap. If you want to see the corresponding code, copy the `FormBootstrap4` directory from the companion disc to a convenient location and look at the contents of `app.component.ts`.

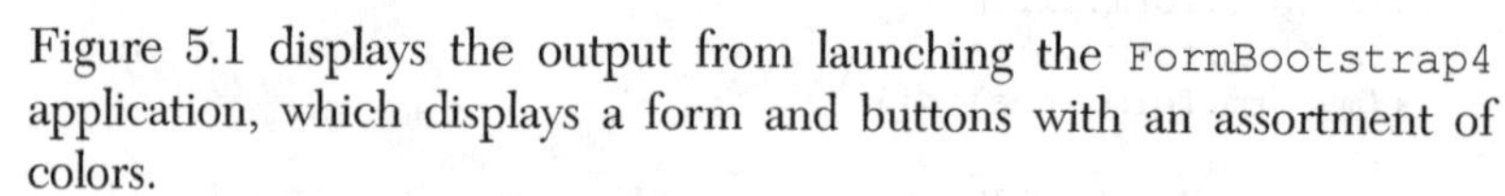

Figure 5.1 displays the output from launching the `FormBootstrap4` application, which displays a form and buttons with an assortment of colors.

This conclude the portion of the chapter regarding `forms` in Angular. The next section discusses Angular Pipes, which provide useful functionality in Angular applications.

Working with Pipes in Angular

Angular supports something called a `pipe` (somewhat analogous to the Unix pipe "|" command), which enables you to filter data based on

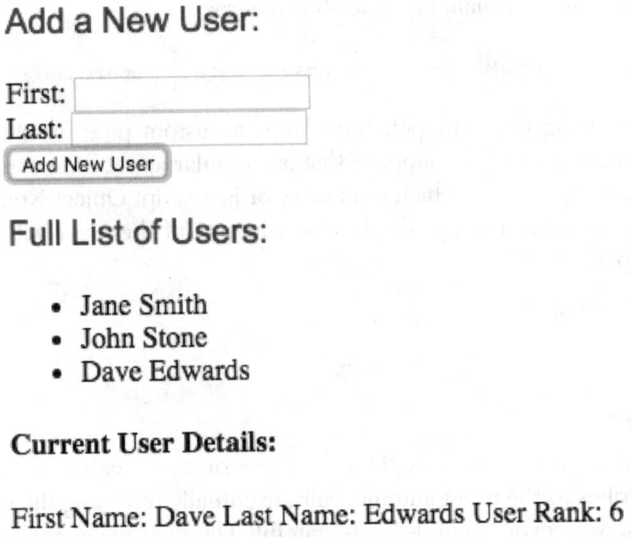

Add a New User:

First:
Last:
Add New User

Full List of Users:

- Jane Smith
- John Stone
- Dave Edwards

Current User Details:

First Name: Dave Last Name: Edwards User Rank: 6

FIGURE 5.1 An angular application with Bootstrap 4.

conditional logic (specified by you). Angular supports built-in pipes, asynchronous pipes, and support for custom pipes. The next two sections show you some examples of built-in pipes, followed by a description of asynchronous pipes. A separate section shows you how to define a custom Angular pipe.

Working with Built-In Pipes

Angular supports various built-in pipes, such as `DatePipe`, `UpperCasePipe`, `LowerCasePipe`, `CurrencyPipe`, and `PercentPipe`. Each of these intuitively named pipes provides the functionality that you would expect: The `DatePipe` supports date values, the `UpperCasePipe` converts strings to uppercase, and so forth.

As a simple example, suppose that the variable food has the value `pizza`. Then the following code snippet displays the string `PIZZA`:

```
<p>I eat too much {{ food | UppercasePipe }} </p>
```

You can also parameterize some Angular pipes, an example of which is shown here:

```
<p>My brother's birthday is {{ birthday | date:"MM/dd/yy" }}
                                                        </p>
```

In fact, you can even chain pipes, as shown here:

```
My brother's birthday is {{ birthday | date | uppercase}}
```

In the preceding code snippet, birthday is a custom pipe (written by you). As another example, suppose that an Angular application contains the variable `employees`, which is an array of JavaScript Object Notation (JSON)-based data. You can display the contents of the array with this code snippet:

```
<div>{{employees | json }}</div>
```

The AsyncPipe

The Angular `AsyncPipe` accepts a `Promise` or `Observable` as input and subscribes to the input automatically, eventually returning the emitted values. Moreover, `AsyncPipe` is stateful: The pipe maintains a subscription to the input `Observable` and keeps delivering values from that Observable as they arrive.

The following code block gives you an idea of how to display stock quotes, where the variable `quotes$` is an `Observable`:

```
@Component({
  selector: 'stock-quotes',
  template: `
    <h2>Your Stock Quotes</h2>
    <p>Message: {{ quotes$ | async }}</p>
  `
})
```

Keep in mind that the `AsyncPipe` provides two advantages. First, `AsyncPipe` reduces boilerplate code. Second, there is no need to subscribe or to unsubscribe from an `Observable` (and the latter can help avoid memory leaks).

One other point: Angular does not provide pipes for filtering or sorting lists (i.e., there is no `FilterPipe` or `OrderByPipe`) because both can be compute intensive, which would adversely affect the perceived performance of an application.

The code sample in the next section shows you how to create a custom pipe that displays a filtered list of users based on conditional logic that is defined in custom code.

Creating a Custom Angular Pipe

Copy the directory `SimplePipe` from the companion disc into a convenient location. Listing 5.7 displays the contents of `app.component.ts` that illustrates how to define and use a custom pipe in an Angular application that displays a subset of a hard-coded list of users.

LISTING 5.7: app.component.ts

```
import { Component } from '@angular/core';
import {User}         from './user.component';
import {MyPipe}       from './pipe.component';

@Component({
  selector: 'app-root',
  template: `
    <div>
      <h2>Complete List of Users:</h2>
      <ul>
       <li
       *ngFor="let user of userList"
         (mouseover)='mouseEvent(user)'
         [class.chosen]="isSelected(user)">
         {{user.fname}}-{{user.lname}}<br/>
       </li>
      </ul>

      <h2>Filtered List of Users:</h2>
      <ul>
       <li
       *ngFor="let user of userList|MyPipe"
         (mouseover)='mouseEvent(user)'
         [class.chosen]="isSelected(user)">
         {{user.fname}}-{{user.lname}}<br/>
       </li>
      </ul>
    </div>
  `
})
export class AppComponent {
  user:User;
  currentUser:User;
  userList:User[];

  mouseEvent(user:User) {
     console.log("current user: "+user.fname+" "+user.lname);
     this.currentUser = user;
  }
```

```
    isSelected(user: User): boolean {
      if (!user || !this.currentUser) {
        return false;
      }

      return user.lname === this.currentUser.lname;
    //return true;
    }

    constructor() {
       this.userList = [
                      new User('Jane','Smith'),
                      new User('John','Stone'),
                      new User('Dave','Jones'),
                      new User('Rick','Heard'),
                     ]
    }
  }
```

Listing 5.7 references a `User` custom component and a `MyPipe` custom component, where the latter is specified in the array of values for the `pipes` property. The `template` property displays two unordered lists of user names. The first list displays the complete list, and when users hover (with their mouse) over a user in the first list, the current user is set equal to that user via the code in the `mouseEvent()` method (defined in the `AppComponent` class). Note that the constructor in the `AppComponent` class initializes the `userList` array with a set of users, each of which is an instance of the `User` custom component.

The second list displays a filtered list of users based on the conditional logic in the custom pipe called `MyPipe`. Listing 5.8 displays the contents of `pipe.component.ts` that defines the pipe `MyPipe` that is referenced in Listing 5.7.

LISTING 5.8: pipe.component.ts

```
import {Component} from '@angular/core';
import {Pipe}      from '@angular/core';

@Pipe({
  name: "MyPipe"
})
export class MyPipe {
  transform(item) {
    return item.filter((item) => item.fname.startsWith("J"));
  //return item.filter((item) => item.lname.endsWith("th"));
```

```
    //return item.filter((item) => item.lname.contains("n"));
  }
}
```

Listing 5.8 contains the `MyPipe` class that contains the `transform()` method. There are three examples of how to define the behavior of the pipe, the first of which returns the users whose first name starts with an uppercase `J` (somewhat contrived, but nevertheless illustrative of pipe-related functionality).

Listing 5.9 displays the contents of the custom component `user.component.ts` for creating `User` instances, which is also referenced via an `import` statement in `app.component.ts`.

LISTING 5.9: user.component.ts

```
import {Component} from '@angular/core';

@Component({
  selector: 'my-user',
  template: '<h1></h1>'
})
export class User {
  fname: string;
  lname: string;

  constructor(fname:string, lname:string) {
     this.fname = fname;
     this.lname = lname;
  }
}
```

The contents of Listing 5.9 are straightforward: there is a `User` class comprising the fields `fname` and `lname` for the first name and last name, respectively, for each new user.

Finally, we need to update the contents of `app.module.ts`, as shown in Listing 5.10, where the modified contents are shown in bold.

LISTING 5.10: app.module.ts

```
import { NgModule }       from '@angular/core';
import { BrowserModule } from '@angular/platform-browser';
import { AppComponent }   from './app.component';
import { MyPipe }         from './pipe.component';
import { User }           from './user.component';
```

```
@NgModule({
  imports:       [ BrowserModule ],
  declarations: [ AppComponent, MyPipe, User ],
  bootstrap:     [ AppComponent ]
})
export class AppModule { }
```

As you can see, Listing 5.10 contains two new `import` statements so that the custom components `MyPipe` and `User` can be referenced in the declarations property.

Launch the application and after a few moments you will see the following in a browser session:

```
Complete List of Users:
```

- Jane-Smith
- John-Stone
- Dave-Jones
- Rick-Heard

```
Filtered List of Users:
```

- Jane-Smith
- John-Stone

Figure 5.2 displays the output from launching this Angular application and filtering based on the users whose first name starts with the capital letter `J`.

FIGURE 5.2 Filtered user list via an Angular pipe.

Now that you understand how to define a basic `Pipe` in Angular, you can experiment with custom pipes that receive data asynchronously. This type of functionality can be very useful when you need to display data (such as a list or a table) after it's updated, without the need for "polling" the source of the data.

Additional information regarding Angular pipes is located here:

https://angular.io/docs/ts/latest/guide/pipes.html

This concludes the portion of the chapter regarding `Pipes` in Angular. The next section discusses Services in Angular applications.

What Are Angular Services?

Sometimes the front-end of a Web application contains presentation logic and some business logic. Angular components comprise the presentation tier and services belong to the business-logic tier. Define your Angular services in such a way that they are decoupled from the presentation tier.

Angular services are classes that implement some business logic, and are designed so that they can be used by components, models, and other services. In other words, services can be providers for other parts of an application.

Because of the Dependency Injection (DI) mechanism in Angular, services can be invoked in other sections of an Angular application. Moreover, Angular ensures that services are singletons, which means that each service consumer will access the same instance of the service class.

A sample Angular custom service is shown here:

```
@Injectable()
export class UpperCaseService {
  public upper(message: string): string {
    return message.toUpperCase();
  }
}
```

The simple service `UpperCaseService` provides one method that takes a string as an argument and returns the uppercase version of that string. The `@Injectable()` decorator is required so that this class can be injected as a dependency. Although this decorator is not mandatory in all cases, it's a good idea to mark your services in this manner. Use the `@Injectable` decorator only when a service (or class) "receives" an injection.

The following is an example of an `app.component.ts` class that invokes the method in the preceding service:

```
import {UpperCaseService} from "./path/to/service/
UpperCaseService";

@Component({
  selector: "convert",
  template: "<button (click)='greet()'>Greet</button>";
})
export class UpperComponent {
  // inject the custom service in the constructor
  constructor(private upperCaseService: UpperCaseService {
  }

  // invoke the method in the uppercaseService class
  public greet(): void {
    alert(this.upperCaseService.upper("Hello world"));
  }
}
```

The preceding code block imports the `UpperCaseService` class (shown in bold) via an `import` statement and then injects an instance of this class into the constructor of the `UpperComponent` class. Next, the `template` property contains a `<button>` element with a click handler that invokes the `greet()` method defined in the preceding code block. The `greet()` method displays an alert whose contents are the result of invoking the `upper()` method in the custom `UpperCaseService` class.

Built-In Angular Services

Angular supports various built-in services, which are organized in different modules. For example, the `http` module (in `@angular/http`) contains support for HTTP requests that involve typical verbs, such as `GET`, `POST`, `PUT`, and `DELETE`. You saw examples of HTTP-based requests in Chapter 4. The routing module (in `@angular/router`) provides routing support, which includes HTML5 and hash routing. The form module (in `@angular/forms`) provides form-related services. Check the Angular documentation for a complete list of built-in services.

An Angular Service Example

Copy the directory `ServiceExample` from the companion disc into a convenient location. Listing 5.13 displays the contents of `app.component.ts` that contains an example of defining a service.

LISTING 5.13: app.component.ts

```
import {Component}  from '@angular/core';
import {Injectable} from '@angular/core';

@Injectable()
class Service {
  somedata = ["one", "two", "three"];
  constructor() { }

  getData()  { return this.somedata; }
  toString() { return "From toString"; }
}

@Component({
  selector: 'app-root',
  providers: [ Service ],
  template: 'Here is the data: {{ service.getData() }}'
})
export class AppComponent {
  constructor(public service: Service) { }
}
```

Listing 5.13 contains a `Service` class that is preceded by the `@Injectable` decorator, which enables us to inject an instance of the `Service` class in the constructor of the `AppComponent` class in Listing 5.13.

A Service with an EventEmitter

This section contains a code sample that uses `EventEmitters` for communicating between a component and its child component.

Copy the directory `UserServiceEmitter` from the companion disc into a convenient location. Listing 5.14 displays the contents of `user.component.ts` that defines a custom component for an individual user.

LISTING 5.14: user.component.ts

```
import {Component} from '@angular/core';

@Component({
  selector: 'user',
  template: '<h2></h>'
})
export class User {
  fname: string;
  lname: string;
  imageUrl: string;
```

```
   constructor(fname:string, lname:string, imageUrl:string) {
      this.fname = fname;
      this.lname = lname;
      this.imageUrl = imageUrl;
   }
}
```

Listing 5.14 is straightforward: the custom `User` class and a constructor with three arguments that represent the first name, last name, and image URL, respectively, for a user.

Listing 5.15 displays the contents of `user.service.ts` that creates a list of users, where each user has a first name, last name, and an associated `PNG` file.

LISTING 5.15: user.service.ts

```
import {Component} from '@angular/core';
import {User}      from './user.component';

@Component({
  selector: 'user-comp',
  template: '<h2></h2>'
})
export class UserService {
  userList:User[];

  constructor() {
     this.userList = [
                   new User('Jane','Smith','src/app/sample1.
                                                         png'),
                   new User('John','Stone','src/app/sample2.
                                                         png'),
                   new User('Dave','Jones','src/app/sample3.
                                                         png'),
                 ]
  }

  getUserList() {
     return this.userList;
  }
}
```

Listing 5.15 imports the `User` custom component (displayed in Listing 5.16), and then defines the `UserService` custom component that uses the `userList` array to keep track of users. This array is initialized in the constructor, and three new `User` instances are created and populated with data. The `getUserList()` method performs the "service" that returns the `userList` array.

Listing 5.16 displays the contents of `app.component.ts` that references the two preceding custom components and renders user-related information in an unordered list.

LISTING 5.16: app.component.ts

```
import {Component}      from '@angular/core';
import {EventEmitter}   from '@angular/core';
import {UserService}    from './user.service';
import {User}           from './user.component';

@Component({
  selector: 'app-root',
  providers: [User, UserService],
  template: `
     <div class="ui items">
       <user-comp
        *ngFor="let user of userList; let i=index"
          [user]="user"
          (mouseover)='mouseEvent(user)'
          [class.chosen]="isSelected(user)">
          USER {{i+1}}: {{user.fname}}-{{user.lname}}
          <img class="user-image" [src]="user.imageUrl"
               (mouseenter)="mouseEnter(user)"
               width="50" height="50">
       </user-comp>
     </div>
    `
})
export class AppComponent {
  user:User;
  currentUser:User;
  userList:User[];
  onUserSelected: EventEmitter<User>;

  mouseEvent(user:User) {
     console.log("current user: "+user.fname+" "+user.lname);
     this.currentUser = user;
     this.onUserSelected.emit(user);
  }

  mouseEnter(user:User) {
     console.log("image name: "+user.imageUrl);
     alert("Image name: "+user.imageUrl);
  }
```

```
  isSelected(user: User): boolean {
    if (!user || !this.currentUser) {
      return false;
    }

    return user.lname === this.currentUser.lname;
  //return true;
  }

  constructor(userService:UserService) {
     this.onUserSelected = new EventEmitter();
     this.userList = userService.getUserList();
  }
}
```

Listing 5.16 contains a `template` property that displays the current list of users (i.e., the three users that are initialized in the constructor in Listing 5.15). Notice the syntax to display information about each user in the list of users:

```
USER {{i+1}}: {{user.fname}}-{{user.lname}}
<img class="user-image" [src]="user.imageUrl"
     (mouseenter)="mouseEnter(user)"
     width="50" height="50">
```

When users move their mouse over the displayed list, the `mouseEvent()` method is invoked to set `currentUser` to refer to the current user. In addition, when users move their mouse over one of the images, the `mouseEnter()` method is invoked, which displays a message via `console.log()` and also displays an alert.

Listing 5.17 displays the contents of `app.module.ts` that references the custom component and custom service.

LISTING 5.17: app.module.ts

```
import { NgModule }      from '@angular/core';
import {CUSTOM_ELEMENTS_SCHEMA} from '@angular/core';
import { BrowserModule } from '@angular/platform-browser';
import { AppComponent }  from './app.component';
import { UserService }   from './user.service';

@NgModule({
  imports:      [ BrowserModule ],
  providers:    [ UserService ],
  declarations: [ AppComponent ],
  bootstrap:    [ AppComponent ],
```

```
  schemas:        [CUSTOM_ELEMENTS_SCHEMA]
})
export class AppModule { }
```

Listing 5.17 has essentially the same contents as the example in Chapter 2 that contains the `schemas` property. The lines shown in bold are the modifications that are required for the code sample in this section.

Displaying GitHub User Data

This section shows you how to read a `Github` user name from an input field, search for that user in `Github`, and then append some details about that user in a list.

Copy the directory `GithubUsersForm` from the companion disc into a convenient location. Listing 5.18 displays the contents of `app.component.ts` that illustrates how to make an `HTTP GET` request to retrieve information about `Github` users.

LISTING 5.18: app.component.ts

```
import { Component }     from '@angular/core';
import { Inject }        from '@angular/core';
import { Http }          from '@angular/http';
import { UserComponent } from './user.component';

@Component({
  selector: 'app-root',
  template: '
    <div>
      <form>
        <h3>Search Github For User:</h3>
        <div class="field">
          <label for="guser">Github Id</label>
          <input type="text" #guser>
        </div>

        <button (click)="findGithubUser(guser)">
          >>> Find Github User <<<
        </button>
      </form>

      <div id="container">
       <div class="onerow">
        <h3>List of Users:</h3>
        <ul>
```

```
          <li *ngFor="let user of users"
              (mouseover)="currUser(user)">
           {{user.field1}} {{user.field2}}</li>
         </ul>
        </div>
      </div>
    </div>
   `
})
export class AppComponent {
  currentUser:UserComponent = new UserComponent('ABC',
                                                 'DEF', '');
  users: UserComponent[];
  githubUserInfo:String = "";
  githubUserJSON:JSON;
  user:UserComponent;
  userStr:String = "";
  guserStr:String = "";

  constructor(@Inject(Http) public http:Http) {
    this.users = [
      new UserComponent('Jane', 'Smith', ''),
      new UserComponent('John', 'Stone', ''),
    ];
  }

  currUser(user) {
    console.log("fname: "+user.field1+" lname: "+user.field2);
    this.currentUser = new UserComponent(user.field1,
                                         user.field2,
                                         user.field3);
  }

  findGithubUser(guser: HTMLInputElement): boolean {
    if((guser.value == undefined) || (guser.value == "")) {
       alert("Please enter a user name");
       return;
    }

    // guser.value is not available in the 'subscribe' method
    this.guserStr = guser.value;

      this.http.get('https://api.github.com/
   users/'+guser.value)

      .map(res => res.json())
      .subscribe(data => {
            //console.log("user = "+JSON.stringify(data));
            this.githubUserInfo = data;
            this.user = new UserComponent(data.name,
```

```
                                this.guserStr,
                                data.created_at);
               this.users.push(this.user); },
          err => {
             console.log("Lookup error: "+err);
             alert("Lookup error: "+err);
          }
        );

     // reset the input field to an empty string
     guser.value = "";

     // prevent a page reload:
     return false;
  }
}
```

Listing 5.18 contains the usual `import` statements, followed by an `@Component` decorator that contains the usual selector property and an extensive template property.

The template property consists of a top-level `<div>` element that contains a `<form>` element and another `<div>` element. The `<form>` element contains an `<input>` element where users can enter a `Github` user name, whereas the `<div>` element contains a `<ul>` element that in turn renders the list of current users. Notice that each `<li>` element in the `<ul>` element handles a `mouseover` event by setting the current user to the element that users have highlighted with their mouse.

The next portion of Listing 5.18 is the definition of the exported class `AppComponent`, which initializes some instance variables, followed by a constructor that initializes the `users` array with two hard-coded users. Next, the `currUser()` method "points" to the user that users have highlighted with their mouse. This functionality is not essential, but it's available if you need to keep track of the current highlighted user.

The `findGithubUser()` method displays an alert if the `<input>` element is empty (which prevents a redundant invocation of the `http()` method). If a user is entered in the `<input>` element, the code invokes an `HTTP GET` request from the `Github` website and appends the new user (as an instance of the `UserComponent` class) to the users array. In addition, an alert is displayed if there is no `Github` that matches the input string.

Another small but important detail is the following code snippet, which keeps track of the user-specified input string:

```
this.guserStr = guser.value;
```

The preceding snippet is required because of the context change that occurs inside the invocation of the `get()` method, which loses the reference to the `guser` argument.

Listing 5.19 displays the contents of `user.component.ts` that contains three strings for keeping track of three user-related fields.

LISTING 5.19: user.component.ts

```
import {Component} from '@angular/core';

@Component({
  selector: 'current-user',
  template: '<h1></h1>'
})
export class UserComponent {
  field1:string = "";
  field2:string = "";
  field3:string = "";

  constructor(field1:string, field2:string, field3:string) {
     this.field1 = field1;
     this.field2 = field2;
     this.field3 = field3;
  }
}
```

Listing 5.19 contains the string properties `field1`, `field2`, and `field3` for keeping track of three attributes from the `JSON`-based string of information for a `Github` user. The property names in the `UserComponent` class are generic so that you can store different properties from the `JSON` string, such as `followers`, `following`, and `created_at`.

You now have a starting point for displaying additional details regarding a user, and you can improve the display by using Bootstrap or some other toolkit for UI-related layouts.

Figure 5.3 displays the output from launching this Angular application and adding information about `Github` users. One thing to notice is that duplicates are allowed in the current sample (the code for preventing duplicates is an exercise for you).

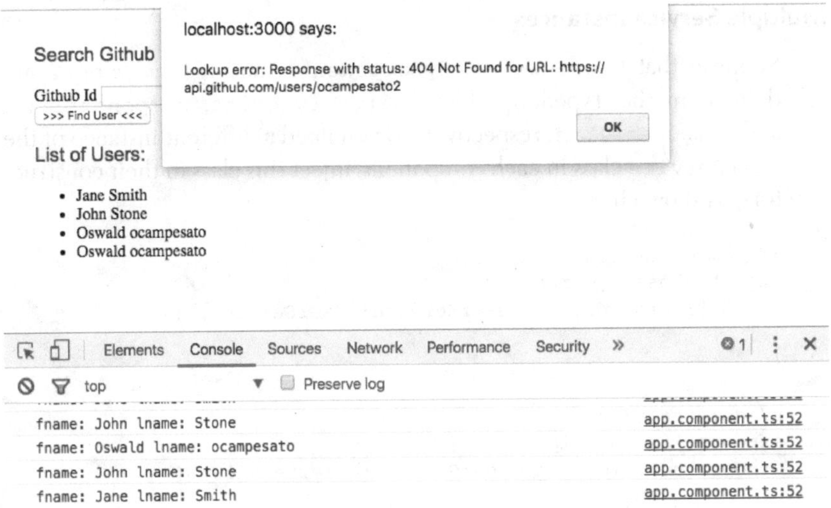

FIGURE 5.3 Search and display GitHub users in a list.

Other Service-Related Use Cases

As you saw in the previous section, services are useful for retrieving external data. In addition, there are other situations that involve sharing data and services in an Angular application. In particular, one application might need multiple instances of a service class, whereas another application might need to enforce a single instance of a service class. Yet another situation involves sharing data between components in an Angular application.

These three scenarios are discussed briefly in the following subsections, and they are based on a very simple `UserService` class that is defined as follows:

```
export class UserService {
   private users: string[];

   adduser(user: string) {
      this.users.push(user);
   }

   getUsers() {
      return this.users;
   }
}
```

Multiple Service Instances

Suppose that `UserService`, `MyComponent1`, and `MyComponent2` are defined in the TypeScript files `user.service.ts`, `component1.ts`, and `component2.ts`, respectively. If you need a different instance of the `UserService` class in each component, inject this class in their constructors, as shown here:

```
// component1.ts
export class MyComponent1 {
  constructor(private userService: UserService) {
  }
}

// component2.ts
export class MyComponent2 {
  constructor(private userService: UserService) {
  }
}
```

In the preceding code, the instance of the `UserService` class in `MyComponent1` is different from the instance of the `UserService` class in `MyComponent2`.

Single Service Instance

Consider the situation in which two Angular components must share the same instance of the `UserService` class. For simplicity, let's assume that the two components are children of the root component. In this scenario, perform the following sequence of steps:

Create a new service component (`ng g s service`).

Include `UserService` in the `providers` array in `app.module.ts`.

Import `MyComponent1` and `MyComponent2` in service.component.ts.

Remove the `UserService` class from the `providers` array in `MyComponent1`.

Remove the `UserService` class from the `providers` array in `MyComponent2`.

Step 2 ensures that the `UserService` class is available to *all* components in this Angular application, and there is only one instance of the `UserService` class throughout the application.

Services and Intercomponent Communication

There are three steps required to send a new user from `MyComponent1` to `MyComponent2`.

Step 1: Define a variable `sendUser` that is an instance of `EventEmitter` and a `sendNewUser()` method in `UserService`:

```
export class UserService {
   sendUser = new EventEmitter<string>();
   ...
   sendNewUser(user:string) {
      this.sendUser.emit(user);
   }
}
```

Step 2: Define an `onSend()` method in `MyComponent1` to send a new user to `MyComponent2`:

```
onSend(user:string) {
   this.userService.sendNewUser(user);
}
```

Step 3: Define an `Observable` in `MyComponent2` to "listen" for data that is emitted from `MyComponent1`:

```
ngOnInit() {
  this.userService.subscribe(...);
}
```

Another way to summarize the logical flow in the preceding code blocks is shown here:

- Users click a button to add a new user.
- The `UserService` instance sends the data to `Component1`.
- The `Component1` instance "emits" the new user.
- The `Component2` instance "listens" for the new user via an `Observable`.

Injecting Services into Services

You have seen how to use DI to inject a service into a component via its constructor. You can also inject services into other services. To do so, use the `@Injectable` decorator in the "injected service":

```
@Injectable
@Component({
```

```
})
export MyService(...)
```

NOTE *DI in Angular only works in classes that have a suitable decorator as part of the class definition.*

Summary

This chapter showed you how to create Angular applications with HTML5 `Forms` as well as `Forms` that contain Angular `Controls` and `FormGroups`. You also saw how to save form-based data in local storage. Next you learned about Angular `Pipes`, along with an example that showed you how to implement this functionality.

You also learned about Angular `Services`, and saw an example that illustrated how to use `Services`. Finally, you learned how use the `http()` method (which returns an `Observable`) of the `Http` class to retrieve data for any `Github` user and display portions of that data in a list of users.

CHAPTER 6

ANGULAR AND EXPRESS

This chapter contains Angular applications that work in conjunction with server-side technologies that are often referred to as the "Node stack." You will learn how to create Angular applications that use `Express` on the server and the NoSQL database Mongo. You will also get an introduction to microservices in this chapter. However, this chapter does not cover the introductory-level material for the Node itself, which is a prerequisite for this chapter (many online tutorials are available).

The first section shows you how to set up a simple JavaScript Object Notation (JSON)-based server (called `json-server`) that you can use with many client-side applications (not just Angular). The second section introduces you to basic `Express`-based applications that can process client-side data requests.

The third section takes a slight detour by introducing you to microservices, along with an Angular application that contains the `forkJoin()` method to simulate a very simple example of microservices. The server-side code consists of three `Express` applications that listen for client-side requests on three different ports.

The fourth section combines an `Express`-based application and an Angular application that issues a data request to the `Express` application. The data request retrieves a set of users from a `Mongo` instance and then returns that list of users to the Angular application. In addition, this code sample uses a technique that you saw in Chapter 4 for returning `JSON`-based data from the server that is converted to a `Promise`, which in turn is converted into an `Observable` that is processed inside the Angular application.

A Minimalistic Node Application

In Chapter 4 you learned how to create Angular applications that can issue HTTP GET and HTTP POST requests to a file server using json-server. The advantage of using json-server is "zero configuration," which is convenient for creating prototypes and demos. In this section you will see how to create a minimalistic code sample using Express, which requires only the file server.js.

Copy the directory SimpleNode from the companion disc to a convenient location. This directory contains all the files for this section. However, if you want to perform the steps yourself, the next section explains how to do so. If you want to use the existing code, skip the following setup section.

Set Up a Node Environment (Optional)

After installing NodeJS on your machine, navigate to a convenient directory and enter the following command:

```
npm init -y
```

The second step is to install Express (which we need in the next section) with the following command:

```
npm install express -save
```

After the preceding command has completed its execution, the current directory will contain the file package.json whose contents are similar to the following:

```
{
  "name": "application-name",
  "version": "0.0.1",
  "private": true,
  "scripts": {
    "start": "node server.js"
  },
  "dependencies": {
    "express": "^4.13.4"
  }
}
```

Create an Express Application

In case you haven't already done so, copy the directory `SimpleNode` from the companion disc to a convenient location. Listing 6.1 displays the contents of `server.js`, which is a simple `Express`-based application.

LISTING 6.1: server.js

```
var express = require("express");
var app = express();

// send a message for the root path ("/"):
app.get('/', function(req, res){
    res.send('hello world');
});

// send a message for the path /pasta:
app.get('/pasta', function(req, res){
    res.send('hello pasta');
});

console.log("Listening on port 3000");
app.listen(3000);
```

Listing 6.1 contains a `require` statement and then a code snippet to initialize the `app` variable. There are two defined routes: The first route is the default route "/" and the second route is "`/pasta`," which displays a corresponding message when users navigate to the `URL` `localhost:3000/pasta`.

Launch the Express Application

Before launching the application, you must install the Node-related dependencies with this command:

```
npm install
```

When the preceding step has completed, invoke the following command:

```
npm start
```

This command will display the following output:

```
Listening on port 3000
```

Now, launch a browser session and navigate to `localhost:3000/pasta`, and if everything was set up correctly, you will see the string `hello pasta` in your browser.

An Application with Angular and Express

In Chapter 4 you saw how to use the `forkJoin()` method in the `Observable` class to issue multiple `HTTP` requests concurrently and then aggregate the results of those requests. This section shows you how replace the `json-server` file server in Chapter 4 with an `Express`-based application, and how to aggregate multiple responses from that Express application. (In a later section you will see how to serve data from multiple Express-based applications.)

The code sample has two parts: The first part shows you how to create a `Node`-based application that returns JSON-based data about individual students (based on the student `id` that is sent to the server). The second part contains an Angular application that uses the `forkJoin()` method to send multiple concurrent requests to the server.

Hence, you need two command shells to do the following:

```
invoke node server.js (from NodeForkJoin/server)
invoke ng serve       (from NodeForkJoin/src)
```

Starting the Server and the Angular Application

Copy the directory called `NodeForkJoin` from the companion disc into a convenient location. Let's start by launching the Express-based application, and then we'll launch the Angular application.

Navigate to the `src/server` directory that contains the file `package.json` (discussed in the next section) and `server.js`. Perform the following step to install the necessary code:

```
npm install
```

Now launch the Express-based application with this command:

```
node server.js
```

This command displays the following output:

```
Listening on port 3000
```

Next, launch the Angular application by opening another command shell, navigating to the `NodeForkJoin/src` directory, and typing the following command:

```
ng serve
```

Navigate to the URL `localhost:4200;` if everything is working correctly, you will see the following text:

Angular HTTP and Observables

A List of Students

- John Smith
- Dave Stone
- Miko Mason

Now that the Angular application is running correctly, the next two sections examine the files `package.json` and `server.js` for the Express-based application, followed by a section that discusses the Angular code.

The Server Code: package.json

Listing 6.2 displays the contents of `package.json` that contains a single dependency that is required for the `Node` application for this section.

LISTING 6.2: package.json

```
{
  "name": "application-name",
  "version": "0.0.1",
  "private": true,
  "scripts": {
    "start": "node server.js"
  },
  "dependencies": {
    "express": "^4.13.4"
  }
}
```

The Server Code: server.js

Listing 6.3 displays the contents of `server.js`, which is the `Node` application for this section.

LISTING 6.3: server.js

```
var express = require("express");
var app = express();

var students = {
                    "1100": {"fname": "John", "lname":"Smith"},
                    "1200": {"fname": "Jane", "lname":"Jones"},
                    "1300": {"fname": "Dave", "lname":"Stone"},
                    "1400": {"fname": "Miko", "lname":"Mason"},
                    "1500": {"fname": "Yuki", "lname":"Smith"}
                }

// add CORS support to allow cross-domain requests
app.use(function(req, res, next) {
  res.header("Access-Control-Allow-Origin", "*");
  res.header("Access-Control-Allow-Headers", "Origin,
                      X-Requested-With, Content-Type, Accept");
  next();
});

// send a message for the root path ("/"):
app.get('/', function(req, res){
    res.send('hello world');
});

// get individual student:
app.get('/json/:id', function(req, res){
    studentid = req.params.id;
    res.send(students[studentid]);
});

console.log("Listening on port 3000");
app.listen(3000);
```

Listing 6.3 contains the `students` array with `JSON`-based data for convenience. In a realistic application, you would retrieve such data from a `Mongo` database (or some other data store).

Listing 6.3 contains three routes, the first of which is the "`/`" default route, which returns a `hello world` message. The second route handles requests that contain an `id` value, which is an exercise for you.

The third route also supports an `id` parameter that is used as an index for the `students` array. The information about the user that is associated with the specified `id` is returned to the client with this code snippet:

```
res.send(students[studentid]);
```

As you can see, Listing 6.3 serves an individual request. The next section shows you how to issue multiple concurrent requests to retrieve student-related information.

The Angular Code

Listing 6.4 displays the contents of `app.component.ts` that queries the `Express`-based server for multiple students and then displays the first name and last name of those students.

LISTING 6.4: app.component.ts

```
import { Component }      from '@angular/core';
import { Http }           from '@angular/http';
import { Observable }     from 'rxjs/Observable';
import 'rxjs/Rx';
import 'rxjs/add/operator/map'

@Component({
  selector: 'app-root',
  template:'
  <h2>Angular HTTP and Observables</h2>
  <h3>A List of Students</h3>
  <ul>
    <li *ngFor="let student of students">
      {{student.fname}} {{student.lname}}
    </li>
  </ul>
  `
})
export class AppComponent {
  students = [];
  studentids:number[] = [1100,1300,1400];

  constructor(private http:Http) { }

  ngOnInit() {
    this.getStudents();
  }
```

```
    getStudents() {
      // map them into a array of observables and forkJoin
      Observable.forkJoin(
        this.studentids.map(
          id => this.http.get('http://localhost:3000/json/'+id)
                    .map(res => res.json())
        ))
        .subscribe(data => {
          this.students = data;
          console.log("subscribe students = "+data);
        })
    }
}
```

Listing 6.4 contains a `template` property that displays an unordered list of student-related information by iterating through the `students` array. The `AppComponent` class in Listing 6.4 initializes `students` as an empty array and `studentids` as an array with four integers. The `AppComponent` class also contains a constructor that initializes an instance of the `Http` class.

Notice the `ngOnInit()` lifecycle method. This method invokes the `getStudents()` method, which is also defined in the `AppComponent` class.

Next, the `getStudents()` method invokes the `forkJoin` method of the `Observable` class to aggregate the result of issuing multiple concurrent `HTTP GET` requests. These multiple requests are made by using the `map()` method to iterate through the values in the `studentids` array. The last line of code in the `getStudents()` method invokes the `subscribe()` method to initialize the contents of the `students` array with the data retrieved from the server. In this example, the `studentids` array contains three values, and therefore the `students` array will contain `JSON`-based data for three students.

The next section contains a generalized version of the code sample in this section: You will learn how to create an Angular application that makes HTTP requests to multiple endpoints to retrieve and aggregate data from those endpoints.

Concurrent Requests and Angular (Version 2)

The code sample in the previous section creates a Node application that serves data requests from an Angular application via the `forkJoin()` method. This method makes multiple concurrent requests and then aggregates the responses. This example also launches `Express` applications that listen on three different port numbers: `4500`, `5500`, and `6500`.

Copy the directory NodeForkJoin2 from the companion disc into a convenient location so that you can view the code that is discussed in the following subsections.

How to Start the Express Server Code

The NodeForkJoin2 directory contains the subdirectory server, which includes a configuration file package.json, four simple Express applications, and a simple shell script (for your convenience) for launching the Express applications that will serve data to the Angular application. The list of files in the server subdirectory is as follows:

```
package.json
server.js
server4500.js
server5500.js
server6500.js
start.sh
```

The file server.js is a generic Express application that responds to requests from a given port number. In addition, server.js was copied to the three JS files server4500.js, server5500.js, and server6500.js. Note that each of these JS files serves data from their respective port numbers (and nothing more). Thus, the files server4500.js, server5500.js, and server6500.js will "listen" on ports 4500, 5500, and 6500, respectively.

Navigate to the server subdirectory (as you did in the previous section) and launch the following command to install the Express-related code:

```
npm install
```

The shell script start.sh contains node commands that launch the three Express applications, and its contents are shown here:

```
node server4500.js &
node server5500.js &
node server6500.js &
```

You can invoke the three preceding commands manually, or you can invoke the following command to launch the three Express applications in the background:

```
./start.sh
```

Launch a browser and navigate to each of the three preceding port numbers (on `localhost`) to confirm that they are responding correctly.

At this point, you are ready to look at the Angular code for this application, which is discussed in the next section.

The Angular Code

Listing 6.5 displays the contents of the file `app.component.ts` that requests data from each of the three `Express` applications that you launched in the previous section.

LISTING 6.5: app.component.ts

```
import {Component}       from '@angular/core';
import {Http}            from '@angular/http';
import {Observable}      from 'rxjs/Observable';
import 'rxjs/Rx';
import 'rxjs/add/operator/map'

@Component({
  selector: 'app-root',
  template:'
  <h2>Angular HTTP and Observables</h2>
  <h3>A List of Students</h3>
  <ul>
    <li *ngFor="let student of students">
      {{student.fname}} {{student.lname}}
    </li>
  </ul>
  `
})
export class AppComponent {
  students = [];

  studentids:any[] = [{url:"http://localhost:4500", sid:1100}
                      {url:"http://localhost:5500", sid:1300}
                      {url:"http://localhost:6500",
sid:1400}];

  constructor(private http:Http) { }

  ngOnInit() {
    this.getStudents();
  }

  getStudents() {
    // map them into a array of observables and forkJoin
```

```
        Observable.forkJoin(
          this.studentids.map(
            loc => this.http.get(loc.url+'/json/'+loc.sid)
                     .map(res => res.json())
          ))
          .subscribe(data => {
             this.students = data;
             console.log("subscribe students = "+data);
          })
    }
}
```

Listing 6.5 is similar to Listing 6.4, and the modified code is shown in bold (the original code is kept intact for comparison purposes). The key idea is to specify a URL with a port number and a student id for each request. The modified forkJoin() command in the getStudents() method in Listing 6.5 shows you how to construct each URL to issue an HTTP request.

Now consider the following generalization: Specify multiple external Web pages that contain information that you want to aggregate in a Web browser, which is an example of combining Angular and microservices.

The next portion of this chapter contains examples of Angular applications that issue requests to Express-based applications that use MongoDB as a data store as well as Jade templates. If you are unfamiliar with Jade, you can replace Jade with EJS or another templating engine of your preference. Although Jade has been deprecated and replaced by Pug, you will find numerous online code samples that use Jade, and this section will help you understand those code samples. As a side point, if you understand how to use Jade, which many people feel is somewhat difficult, other templating engines will probably be much easier for you to learn.

If you are new to MongoDB, please read an online tutorial that discusses MongoDB. Then you will be ready for the next section, which shows you how to set up an Express-based application that connects to an instance of a MongoDB database.

An Express Application with MongoDB (Part 1)

This section (part 1) explains how to set up an Express application with Mongo, followed by a section (part 2) that discusses the files in the Express application.

Make sure that you have installed `Mongo`, `node`, and `nodemon` on your machine before you attempt to launch the application in this section. Next, copy the directory `ExpressMongoJade` from the companion disc to a convenient location. This directory contains the server-side code that will process requests from the Angular application `NG2NodeApp`, which is discussed later in this section.

Here are the steps that you must perform to launch the existing `Express` application that accesses data in a `MongoDB` database:

1) Install Mongo on your machine.
2) Launch `mongod`.
3) Launch `cd ExpressMongoJade`.
4) Launch mongo load-newusers.js.
5) Launch npm install.
6) Launch npm start.
7) Navigate to `localhost:3000`.

After you have installed Mongo, add the `bin` directory to the `PATH` environment variable in a command shell, and then start the `mongod` daemon process:

```
mongod
```

You can also specify a different directory for the data files:

```
mongod --dbpath <full-path-to-a-directory>
```

Step 4 populates a Mongo database with seed data. Step 5 is required if you do not already have the `node_modules` directory (which exists in the sample code), and Step 6 launches the Express application (by executing the `script bin/www`).

If everything went well, you will see the following text in a browser session:

Express

Welcome to Express

The URL `localhost:3000/userlist` displays the current list of users in the Mongo database, an example of which is shown here:

List of Users

- 0 Janice | Smith
- 1 Steven | Stone

- 2 Yuki | Tanaka
- 3 Hideki | Hiura
- 4 Himiko | Yamamoto

Now let's take a look at the contents of `package.json` (displayed in Listing 6.6), wich lists all the dependencies for this application.

LISTING 6.6: package.json

```
{
  "name": "expressmongojade",
  "version": "0.0.0",
  "private": true,
  "scripts": {
    "start": "node ./bin/www"
  },
  "dependencies": {
    "body-parser": "~1.17.1",
    "cookie-parser": "~1.4.3",
    "debug": "~2.6.3",
    "express": "~4.15.2",
    "jade": "~1.11.0",
    "mongodb": "^2.2.26",
    "monk": "^4.0.0",
    "morgan": "~1.8.1",
    "serve-favicon": "~2.4.2"
  }
}
```

As you can see, Listing 6.6 specifies various Node modules, such as `express`, `jade` (a templating engine), and `mongo`. Note that the version numbers will probably change in some cases, as well as the version of `node` and `npm` on your machine.

The Server-Side Custom Files (Part 2)

Before delving into the contents of various Angular-related files for the project that was introduced in the previous section, let's take a brief tour of the names of some important files and their purpose.

As a starting point, the file `app.js` (in Listing 6.7) contains the `Node` application code. This code contains middleware, a reference to an instance of a `Mongo` database, a reference to the `views` directory, and route definitions in the routes subdirectory. You launch `app.js` via this command:

```
npm start
```

Whenever you issue a valid `HTTP GET` request, that request is handled via a three-step process: The first step involves the appropriate code in the Express application (defined in `app.js`), which invokes an appropriate block of code in `routes/index.js` as the second step, and then data is passed to a `Jade` template as the third step. At the end of this process, an HTML page will be generated and returned to the client that made the initial request.

By convention, the Jade templates for this application are located in the `views` subdirectory. For this example, the Jade templates include `user-list.jade` (for displaying all users) and `index.jade` (for displaying a simple message). Each of these Jade templates manipulates the data that it receives (from a route in `routes/index.js`) to populate a Jade template.

Because the Jade template and the data are *dynamically* combined to generate an HTML Web page, there are no HTML Web pages in the application that correspond to the Jade templates.

One other point: The route that matches `/userlist` is the most complex route because it first extracts the list of all users from a Mongo database and then passes that list of users to the Jade template `views/userlist.jade`.

To summarize the steps in a bullet fashion, the flow of logic works like this:

1) A client-side request arrives and is processed by `app.js`.
2) That request is "routed" to `routes/index.js`.
3) Data is passed to a Jade template in the `views` directory.
4) An HTML Web page is generated.
5) The generated HTML Web page is returned to the client.

The app.js File

Listing 6.7 displays the contents of `app.js` for a `Node`-based application for retrieving user-related data from a Mongo database.

LISTING 6.7: app.js

```
var express = require('express');
var path = require('path');
var favicon = require('serve-favicon');
var logger = require('morgan');
```

```
var cookieParser = require('cookie-parser');
var bodyParser = require('body-parser');

var mongo = require('mongodb');
var monk = require('monk');
var db = monk('localhost:27017/test');

var routes = require('./routes/index');
//var users = require('./routes/users');

var app = express();

// view engine set-up
app.set('views', path.join(__dirname, 'views'));
app.set('view engine', 'jade');

// uncomment after placing your favicon in /public
//app.use(favicon(dirname + '/public/favicon.ico'));
app.use(logger('dev'));
app.use(bodyParser.json());
app.use(bodyParser.urlencoded({ extended: false }));
app.use(cookieParser());
app.use(express.static(path.join(__dirname, 'public')));

// Make our db accessible to our router
app.use(function(req,res,next){
    req.db = db;
    next();
});

app.use('/', routes);
app.use('/userlist', routes);
//app.use('/users', users);

/// catch 404 and forwarding to error handler
app.use(function(req, res, next) {
    var err = new Error('Not Found');
    err.status = 404;
    next(err);
});

// development error handler
// will print stacktrace
if (app.get('env') === 'development') {
    app.use(function(err, req, res, next) {
        res.status(err.status || 500);
        res.render('error', {
            message: err.message,
            error: err
        });
    });
```

```
}

// production error handler
// no stacktraces leaked to user
app.use(function(err, req, res, next) {
    res.status(err.status || 500);
    res.render('error', {
        message: err.message,
        error: {}
    });
});

module.exports = app;
```

Listing 6.7 contains code that tends to be particularly confusing to beginners, partly because of limited documentation (and perhaps the non-intuitiveness of the code). After enough repetition, you will become comfortable with this code, which appears frequently in `Node`-based applications (at least there is some good news for you).

The first part of Listing 6.7 contains standard `require` statements, initializes routes (`routes/index.js`) and users (`routes/users.js`), and then initializes `app` as an `Express` application. Another section of Listing 6.7 sets up the middleware, followed by Mongo-related variables such as `url` (with a default value).

The index.js File

Listing 6.8 displays the contents of `routes/index.js` that defines routes for user-related actions, such as retrieving a list of users from a Mongo instance and saving a new user.

LISTING 6.8: index.js

```
var express = require('express');
var router = express.Router();

// GET request for home page
router.get('/', function(req, res, next) {
  res.render('index', { title: 'Express' });
});

// GET the userlist page
router.get('/userlist', function(req, res) {
    var db = req.db;
```

```
    var collection = db.get('users');
    collection.find({},{},function(e,docs){
        res.render('userlist', {
            "userlist" : docs
        });
    });
});

module.exports = router;
```

Listing 6.8 contains the `router` variable, which manages the default route and two other routes. However, the routes in Listing 6.8 pass JSON-based data to Jade-based templates, which are in the `views` subdirectory. The naming convention makes it easy: The default route passes the value for `title` to the template `views/index.jade` and the second route passes the value for `title` (a different value) to the template `views/hellow-orld.jade`.

The code in the second route performs the following sequence of steps:

1) First, initialize the variable collection as a reference to the users database.
2) Initialize `userlist` as a reference to the documents in the users database.
3) Pass the `userlist` variable to the `views/userlist.jade` template.

As you can see, a lot of work is being performed on your behalf in the third route.

Note that the Jade templates are populated with various bits of information during the process of dynamically generating HTML Web pages that are send back to the browser.

Go back to Listing 6.3 and you will see the following code snippet, which specifies the `Jade` template engine:

```
app.set('view engine', 'jade');
```

One advantage of defining the file `index.js` with the code in Listing 6.4 is that you avoid cluttering the contents of `app.js`, which contains primarily initialization code.

For your convenience, this application contains the optional file `load-newusers.js`, which you can launch from the command line to populate a database with a set of users.

The `userlist.jade` Template

Listing 6.9 displays the contents of `routes/userlist.jade` that takes the list of users as input to populate a template with user-based data.

LISTING 6.9: userlist.jade

```
extends layout

block content
    h1.
        List of Users
    ul
        each user, i in userlist
            li #{i} #{user.fname} | #{user.lname}
```

If you are new to Jade, you might be surprised by the compactness of the code in Listing 6.9. The variable `userlist` (which contains an array of users) is passed in from the route in `routes/index.js` that matches the route `/userlist`. Each `user` in `userlist` is a JSON-based string that contains an `fname` property and an `lname` property.

For your convenience, the following handy website converts HTML to `Jade`, thereby increasing your productivity when you create Jade templates:

http://html2jade.org/

Launching the Server-Side Application

Just to review the necessary steps for launching the application, you need to start the `Mongo` daemon process and launch the `Express` application. Specifically, you need to perform the following two steps:

Step 1: Launch the `Mongo` daemon process: `mongod`.

Step 2: Launch the `Express` application: `npm start`.

One other point: The code sample in this section was created with the `express` command. If you want to create your own Express-based applications, you need to execute the following two commands:

```
sudo npm install -g express-generator
express myexpressappname
```

You can use the files in this section to populate your new Express application, or you can provide your own custom content.

Now that you have completed this section, you are ready to work with an Angular application that accesses data from a Mongo database, which is the topic of the next section.

An Angular Application with Express

This section contains an Angular application that sends data requests to an `Express` application, which in turn sends data from a hard-coded array of values. If you want to retrieve data from a Mongo database, you can incorporate some of the code in the `ExpressMongoJade` application.

Another point: The code in this code sample does not illustrate "best practices," but instead shows you some functionality (such as the code in the `userInfo()` method in Listing 6.10) that might be useful for your own application.

Now copy the directory `NG2NodeApp` from the companion disc to a convenient location. The file `app.component.ts` in Listing 6.10 is very similar to the Angular application in Chapter 4 involving an `HTTP GET` request. We need to make very minor changes to the new version of `app.component.ts`, which includes specifying a URL that matches the code in `app.js` in the previous section.

LISTING 6.10: app.component.ts

```
import { Component }    from '@angular/core';
import {Inject}         from '@angular/core';
import {Http}           from '@angular/http';
import 'rxjs/add/operator/map';

@Component({
   selector: 'app-root',
   template: '<button (click)="httpRequest()">Student
                                            Details</button>
             <div>
               <li *ngFor="let fname of firstNames; let i =
                                                    index;">
                  {{fname}} * {{lastNames[i]}}
               </li>
             </div>
             `
})
export class AppComponent {
  userData:any;
  firstNames:Array<string> = [];
```

```
  lastNames:Array<string>  = [];
  studentIDs:Array<string> = [];

  constructor(@Inject(Http) public http:Http) { }

  httpRequest() {
    this.http.get('http://localhost:3000/students')
      .map(res => res.json())
      .subscribe(
        data => this.userData = data,
        err => console.log('Error: '+err),
        () => this.userInfo()
      );
  }

  userInfo() {
     this.userData = JSON.stringify(this.userData);

     // populate several JavaScript arrays with data
     JSON.parse(this.userData, (key, value) => {
       if(key == "fname") {
          this.firstNames.push(value);
       } else if (key == "lname") {
          this.lastNames.push(value);
       } else {
          this.studentIDs.push(value);
       }
     });
  }
}
```

Listing 6.10 illustrates how easily you can separate the code in an Angular application from the server-side code in a NodeJS application. Notice how the `*ngFor` code block uses the variable `i` to keep track of the current index in the `firstNames` array, and then uses this same variable to find the corresponding last name in the `lastNames` array.

When users click the `<button>` element in the `template` property, the `httpRequest()` method is invoked to retrieve student-related data from an Express application. After the `httpRequest()` method has completed, the `firstNames` array and the `lastNames` array are populated in the `userInfo()` method. The output looks like this:

- John * Smith
- Jane * Jones
- Dave * Stone
- Miko * Mason
- Yuki * Smith

Recall that the `Express` application needs to bypass the `Jade` templates and return the `JSON`-based data to the browser. Listing 6.11 displays the contents of `app.js` (located in the p`NG2NodeApp/server` subdirectory) that returns a hard-coded array of student information.

LISTING 6.11: app.js

```
var express = require("express");
var app = express();

var students = {
                    "1100": {"fname": "John", "lname":"Smith"},
                    "1200": {"fname": "Jane", "lname":"Jones"},
                    "1300": {"fname": "Dave", "lname":"Stone"},
                    "1400": {"fname": "Miko", "lname":"Mason"},
                    "1500": {"fname": "Yuki", "lname":"Smith"}
                  }

// add CORS support to allow cross-domain requests
app.use(function(req, res, next) {
  res.header("Access-Control-Allow-Origin", "*");
  res.header("Access-Control-Allow-Headers", "Origin,
                       X-Requested-With, Content-Type, Accept");
  next();
});

// send a message for the root path ("/"):
app.get('/', function(req, res){
    res.send('hello world');
});

// get all students:
app.get('/students', function(req, res){
    res.send(students);
});

console.log("Listening on port 3000");
app.listen(3000);
```

Listing 6.11 starts with a hard-coded list of users in the `students` array, along with the `/students` route that returns the contents of the `students` array.

Navigate to the `NG2NodeApp/server` subdirectory and invoke the following command:

```
node server.js
```

Next, navigate to the NG2NodeApp/src subdirectory and launch the Angular application with this command:

```
ng serve
```

Click the <button> element; if everything was set up correctly, you will see the following list of students:

- John * Smith
- Jane * Jones
- Dave * Stone
- Miko * Mason
- Yuki * Smith

The next section outlines the steps for retrieving Japanese text that is stored in a Mongo database.

Angular and Japanese Text (Optional)

You can easily modify the code sample ExpressMongoJade in the previous section to work with Japanese text. There are two modifications that you need to make for this code sample:

1) Populate a collection with Japanese text (mongo load-japanese.js).
2) Define a new route /japanesejson in index.js.

Listing 6.12 displays the contents of load-japanese.js that populates a Mongo collection with a simple dictionary of English words and their counterparts in Japanese.

LISTING 6.12: load-japanese.js

```
// you can invoke either of these commands:
// mongo localhost:27017/japanesedb load-japanese.js
// mongo localhost:27017/japanesedb --quiet load-japanese.js

// drop the current database
db.dropDatabase()

// insert data
db.dictionary.insert({ english: 'eat',
                       japanese: 'たべる'});
db.dictionary.insert({ english: 'drink',
                       japanese: 'のむ'});
```

```
db.dictionary.insert({ english: 'return',
                       japanese: 'かえる'});
db.dictionary.insert({ english: 'dance',
                       japanese: 'おどる'});
db.dictionary.insert({ english: 'guess',
                       japanese: 'アングルにわとてもかっこいいで
                                                          すよ！'});
```

Now add the following new route to `index.js`:

```
router.get('/japanesejson', function(req, res) {
    var db = req.db;
    var collection = db.get('dictionary');

    collection.find({},{},function(e,docs){
        res.json(docs);
    });
});
```

The preceding code block is very similar to the modified route in the previous section: Instead of passing data to a `Jade` template, simply return the data via the code snippet shown in bold.

Angular Universal

Universal Angular supports server-side rendering for Angular, and the original home page is located here (but not used in this section):

https://github.com/angular/universal

However, the Angular core team has made very extensive modifications to integrate the code in the preceding link into Angular. The Universal Angular code is integrated in `@angular/platform-server`.

A very rudimentary sample application involving Universal Angular is downloadable here:

https://github.com/robwormald/ng-universal-demo/

Download and uncompress the code from the preceding link in a convenient location and then type the following command:

```
npm install
```

After the preceding command has completed, launch the application:

```
npm start
```

Navigate to the URL `localhost:8000` and you will see the following simple interface:

Universal Demo

Home Lazy

Home View

Click either of the preceding links and you will see some basic text displayed.

The Server File main.server.ts

Listing 6.13 displays the contents of the file `main.server.js` that is executed when you launch the Angular Universal application in the previous section.

LISTING 6.13: main.server.js

```
import 'zone.js/dist/zone-node';
import { platformServer, renderModuleFactory }
         from '@angular/platform-server'
import { enableProdMode } from '@angular/core'
import { AppServerModule } from './app.server'
import { AppServerModuleNgFactory }
         from './ngfactory/src/app.server.ngfactory'
import * as express from 'express';
import {ngExpressEngine} from './express-engine'

enableProdMode();

const app = express();

app.engine('html', ngExpressEngine({
        baseUrl: 'http://localhost:4200',
        bootstrap: [AppServerModuleNgFactory]
}));

app.set('view engine', 'html');
app.set('views', 'src')

app.get('/', (req, res) => {
        res.render('index', {req});
});

app.get('/lazy', (req, res) => {
```

```
        res.render('index', {req});
});

app.listen(8000,() => {
        console.log('listening...')
});
```

Listing 6.13 contains `import` statements for Angular code and for Express-related code that is familiar from code samples in the first part of this chapter. In particular, the `app` variable is initialized as an Express application, followed by some middleware setup steps. Two simple routes—"`/`" and "`/lazy`"— are defined to demonstrate how to handle route-based requests. Note that an application with multiple routes would probably place its definitions in a separate file (located in a separate subdirectory).

Notice that Listing 6.13 specifies the `src` directory as the directory that contains view-related files. In particular, the `/lazy` route is "mapped" to the file `index.html`, which is collocated in the `src` subdirectory.

The Web Page index.html

Listing 6.14 displays the contents of the Web page `index.html` that contains the custom element `<demo-app>`.

LISTING 6.14: index.html

```
<html>
<head>
  <meta charset="UTF-8">
  <title>Universal Test</title>
</head>
<body>
  <demo-app></demo-app>
</body>
</html>
```

Listing 6.14 is straightforward: The `<body>` element contains the `<demo-app>` custom element, which is the top-level element for the Angular application.

The TypeScript File app.ts

The TypeScript file `app.ts` contains route-related code, which was discussed in greater detail in Chapter 5.

Listing 6.15 displays the contents of the `app.ts` file, which is also located in the `src` subdirectory.

LISTING 6.15: app.ts

```
import { NgModule, Component } from '@angular/core'
import { BrowserModule }       from '@angular/
                                          platform-browser'
import { RouterModule }        from '@angular/router'

@Component({
   selector: 'home-view',
   template: '<h3>Home View</h3>'
})
export class HomeView {}

@Component({
   selector: 'demo-app',
   template: `
     <h1>Universal Demo</h1>
     <a routerLink="/">Home</a>
     <a routerLink="/lazy">Lazy</a>
     <router-outlet></router-outlet>
   `
})
export class AppComponent {}

@NgModule({
   imports: [
      BrowserModule.withServerTransition({
        appId: 'universal-demo-app'
      }),
      RouterModule.forRoot([
         { path: '', component: HomeView, pathMatch: 'full'},
         { path: 'lazy', loadChildren: './lazy.
                                      module#LazyModule'}
      ])
   ],
   declarations: [ AppComponent, HomeView ],
   bootstrap: [ AppComponent ]
})
export class AppModule {}
```

Listing 6.15 contains three main parts: the `HomeView` class, the `AppComponent` class, and the `AppModule` class (all of which are exported).

The `HomeView` class is associated with the "home" route, which is the first of the two routes in this sample application.

The `AppComponent` class is preceded by an `@Component` decorator, whose `selector` property specifies the `<demo-app>` custom element. In addition, the `template` property contains the definitions of the two routes.

Finally, the `AppModule` class is preceded with an `@NgModule` decorator, whose contents are typically located in a separate file such as `app.module.ts`.

In essence, Universal Angular applications involve a two-step process: Use `npm` to launch an Express-based application that contains a route to an `index.html` Web page. This Web page contains an element that is the root of an Angular application, and the latter is defined in TypeScript files.

Working with Microservices (Optional)

This section contains a condensed and simplified overview of microservices. You will learn about some of the advantages and the disadvantages of microservices, followed by an example of an Angular application that uses microservices. Although Angular supports Microservices (so it's a relevant topic for this chapter), you can skip this section with no loss of continuity.

Advantages of Microservices

The microservices architecture decomposes monolithic applications into a set of services without reducing the functionality of the original applications. Each service is accessible via its exposed application programming interface (API). The use of microservices simplifies the process of developing, maintaining, and enhancing "one-purpose" services.

Second, microservices allows for the development of services in an independent manner. Each new service can be written using the technology that is most suitable for that service (which means current technology is not mandatory).

A third advantage is that a microservice can be deployed independently of other microservices. Fourth, services can be scaled independently of each other.

Disadvantages of Microservices

A collection of many independent microservices can create complexity. Communication between microservices requires a decision regarding interprocess communication. Moreover, the distributed nature of microservices necessitates handling failures that can arise due to timeouts and other causes.

For example, if transaction-oriented microservices interact with multiple databases, then processing and coordinating those transactions can be complicated. Keep in mind that one of the primary goals of adopting microservices is to reduce the complexity of the interdependencies that can occur in large monolithic applications.

Some people think of microservices as a "fine-grained service-oriented architecture (SOA)." However, SOA requires the Web Services Definition Language, which defines service endpoints and is strongly typed, whereas microservices have very simple connections and smart endpoints. Another important difference is that SOA is often stateless, whereas microservices are stateful and keep data and logic together. In general, SOA is complex and heavyweight, whereas microservices are simpler and lightweight independent processes.

Summary

This chapter showed you how to create simple `Express`-based Node applications that can serve requests from Angular applications. You saw how to use the NoSQL database `MongoDB` to store and retrieve data for Web applications. You also learned how to launch multiple Express applications on different port numbers to serve data to an Angular application. You then saw an overview of microservices, which can work seamlessly with Angular. Finally, you learned some basic concepts about Angular Universal, which involves server-side rendering of Angular code.

CHAPTER 7

FLUX, REDUX, GRAPHQL, APOLLO, AND RELAY

This chapter contains a broad introduction to the JavaScript-based technologies `Flux`, `Redux`, `GraphQL`, `Apollo`, and `Relay`. Except for Apollo (which is under the aegis of Meteor), Facebook developed the other technologies in the preceding list. Although these technologies are used in ReactJS-based Web applications, they can also be used in Angular applications as well (and hence this chapter).

Because you can create Angular applications without any of the material in this chapter, the main purpose is to explain some of the concepts in the preceding technologies. Moreover, there is only one complete code sample in this chapter (see the section that discussed Apollo). Although there are various online code samples that combine Angular and Redux, keep in mind that many of them were written for earlier versions of Angular and they might require some modification to work with Angular 4.0.0.

The first section of this chapter describes the `Flux` architecture, designed by Facebook, which has many implementations (including `Redux` and `Relay`). The `Flux` architecture was initially created for developing client-side Web applications. Because `Flux` is language agnostic, you can use the `Flux` pattern in React-based applications, Angular applications, and others. Note that you will also see the `Flux` architecture described as the `Flux` pattern, perhaps in the same sense that model–view–controller (MVC) is also a pattern.

The second section discusses `Redux`, which is a toolkit whose purpose is to store application state outside the application. Interestingly, this approach for storing application state has some advantages, as you will see later in this chapter. You will see a nice example that illustrates how to use `Redux` in an Angular application for keeping track of items.

The third section describes `GraphQL`, which is a JavaScript toolkit that receives `Relay` requests. `GraphQL` processes those requests by retrieving the matching data from a data store, which can be a relational data store or a `NoSQL` data store. This section contains a basic Angular application that uses `GraphQL`.

The fourth section discusses `Apollo`, which is client-side JavaScript that sends data requests to a `GraphQL` toolkit that resides on a server. Apollo has recently gained traction because of its advantages over `Relay`.

One point to remember is that medium-sized and large Web applications benefit from toolkits such as `Redux` or `GraphQL` more so than small Web applications (where MobX might be more suitable). Consequently, if you are currently working on basic Web applications, portions of this chapter might be optional for you right now.

What Is Flux?

The `Flux` pattern is based on other design patterns (such as `Observer` and `Command Query Responsibility Segregation` [CQRS]). `Flux` provides a good foundation for developing sophisticated applications. The primary purpose of `Flux` is its support for one-way data flow. Watch this video for Flux tips:

https://angular-university.io/course/angular2-ngrx

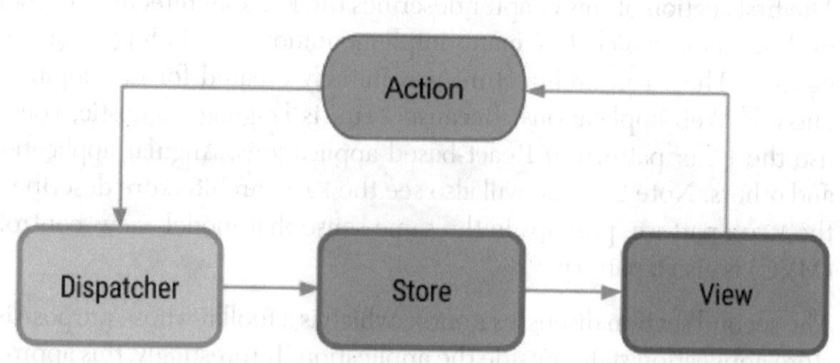

FIGURE 7.1 The elements of the Flux pattern.

The components of the Flux pattern are Action, Dispatcher, and Store. The relationships among these components is depicted here:

```
Action --> Dispatcher --> Store --> View (and back to the
Action)
```

The diagram in Figure 7.1 displays the Flux architecture (copied from the GitHub repository at *https://github.com/lgvalle/android-flux-todo-app*).

Here's a one-paragraph description of the Flux pattern: When users perform a gesture (e.g., click a button) in a Web application, that gesture is converted into an Action that "models" a state change via (1) a *type* that describes the type of action and (2) a *value* that contains a new value. The Action is sent to a Dispatcher, which in turn sends the Action and the current Store to the appropriate reducer function to process the Action. The reducer function contains a switch statement and conditional logic to determine which case statement matches the Action type, after which the code in the matching case statement is executed, and then an updated Store (sometimes the Store does not change) is returned from the reducer function. Next, the View objects that are registered to detect state-related changes will update the contents of the Web page accordingly. This process repeats every time users perform a gesture in the Web application. The more general scenario involves multiple reducers and multiple stores, which means that additional logic is required in the Dispatcher. In addition, an Action can originate from a server as well as a user-initiated event.

NOTE *Application components in a Flux-based Web application do not communicate directly with each other.*

Notice that the preceding flow of data is *always* unidirectional. Moreover, data can *only* be updated in one case statement in a reducer, which in turn simplifies the debugging process.

What Is an Action?

An Action is a JavaScript object that contains JSON-based data, with a mandatory property called type. The purpose of an Action is to specify what needs to be modified in the application state to create a new state. Think of an Action as a "message" containing information or instructions for updating the state of an application (such as adding a new user, deleting a user, and so forth). A key point: Instead of modifying the current state, a new object is created as a result of "applying" an action to the current state.

In brief, an application involves many `Actions` (performed as asynchronous operations): one `Dispatcher`, one or more `Stores`, and one or more `View` components. `Actions` typically contain data that is associated with user-initiated events (such as user input, key strokes, mouse-related events, and so forth) that occur in a `View` component.

Although an `Action` can be a string or an array, an `Action` is often a `JSON`-based object literal. For example, an `Action` with an `ADD` operation involving a new user would look something like this:

```
{ type: 'ADD_USER', value: 'John Smith' }
```

The preceding `Action` is forwarded (via a dispatcher) to a reducer function that adds the new user to an external data store. You will see more details for this use case later in the chapter.

Another point to keep in mind is that the properties of an `Action` can have different names. For example, the following `Action` is equivalent to the earlier code snippet:

```
{ kind: 'ADD_USER', data: 'John Smith' }
```

Note that you will sometimes see a JavaScript "helper" function (called an "Action Creator") that returns an `Action`, an example of which is shown here:

```
function actionCreator(type, value) {
   return { type, value };
}
```

What Is a Reducer?

A `Reducer` is a JavaScript function that has no side effects (also called a "pure function"), and its purpose is to transform the current state of an application into a new state, based on the contents of an `Action` object. In general, one function is associated with each action type. `Reducers` are JavaScript functions that change the state of an application.

The JavaScript `Array` object has a `reduce()` method, an example of which is shown here:

```
function sum(a,b) { return a + b }
[1,2,3,4].reduce(sum,0);
```

`Redux` leverages the concept of the JavaScript `reduce()` method as follows: A `reducer()` method has a `state` argument and an `action` argument, with conditional logic that returns a new state.

The Store and Application State

A `Store` is a container that holds the global state tree (a JavaScript object) and dispatches actions. A store also holds the reducers and provides subscriptions to changes of state.

Keep in mind that the Flux-based Web applications maintain the application state *external* to the application, and that application components do not communicate with each other when the application state changes; instead, they obtain the application state from the contents of the `Store`. Because cross communication among multiple components can introduce bugs that are difficult to find, the Flux architecture removes this source of errors (think of what can happen in an application that contains hundreds of components that have cross-communication).

Because the reducing function is the only place where the `Store` can be updated, you know where to look in the event that a data-related error occurs.

What Is Redux?

`Redux` is an open source toolkit created by Dan Abramov that implements the `Flux` pattern (the latter was created by Facebook); its home page is located here:

https://github.com/reactjs/react-redux

`Redux` consists of `Actions` (for messages), `Reducers` (for changing state), and a `Store` (for holding the global state of an application), and together these components implement the `Flux` pattern.

One simple use case that illustrates the role of `Redux` is a Web application that displays a list of names and also allows users to add new names. The flow of data involves the creation of an action (for adding a user), dispatching the action to a method that does the actual adding of a new user, followed by updating a store with the new state of the application, and finally updating the Web page so that it displays the updated list of names.

DevTools for Redux with hot reloading, action replay, and customizable user interface (UI) is downloadable here:

https://github.com/gaearon/redux-devtools

Data Structures for Application Data

In addition to storing the application state in a store, there is also the question of which data structures to use for application data. Data structures that work well for server-side code or for the view component do not necessarily work best for the structure of the data in the store. In other words, you cannot simply mimic the same data structures (as tempting as this probably is) for the data in the store and expect the application to be performant. One article that discusses this point is located here:

https://hackernoon.com/avoiding-accidental-complexity-when-structuring-your-app-state-6e6d22ad5e2a#.jzmngm3cf

When Should You Use Redux?

Although answers to this question vary, Dan Abramov himself says that people often start using `Redux` too soon. Perhaps the right time to use `Redux` is when the complexity of an application warrants the use of `Redux`. However, this response raises a new question: What is the right complexity?

Fortunately, there are some guidelines that you can follow to make a determination regarding the use of `Redux`. The following link also addresses the use of `Redux`:

http://jamesknelson.com/5-types-react-application-state/

This article is definitely worth the time to read about the details of the preceding numbered list.

There are some additional considerations as well. For example, when you hot reload components, the state is removed from the existing component tree. Hence, if the state of an application is maintained externally, then that state can be reloaded along with the updated components, thereby maintaining a consistent state. Other benefits include better testing facility, centralized logic, time travel debugging, and predictable state updates.

Incidentally, in addition to `Redux`, there are many implementations of `Flux`, so you do have options (so far, `Redux` is the most popular one). One alternative is `Mobx`, which is discussed later in this chapter.

Simple Examples of Reducers

As you will see in this section, the custom code in reducers often contains a `switch` statement with multiple `case` statements (and a default case). The following code block is a generic example of a reducer that takes a `state` argument (initialized as an empty object) followed by an `action` argument:

```
const myReducer = (state = {}, action) => {
  switch (action.type) {
    case 'ADD':      return { ... }
    case 'DELETE': return { ... }
    case 'UPDATE': return { ... }
    default:          return state
  }
}
```

The details of the `case` statements are obviously application specific. The following subsections illustrate reducers whose state is a numeric counter, along with an example where the state is an array of strings. Note that some of the code samples use the "spread" operator in JavaScript.

A Reducer to Add/Subtract Numeric Values

In an HTML Web page, a JavaScript variable can keep track of a numeric counter. Now let's learn how to use a reducer function to keep track of such a counter. The "state" of the counter is simply its current value, which will be "stored" separately from the JavaScript code in the Web page.

Suppose we have a Web page that contains two buttons: one button subtracts 1 from a counter and the other button adds 1 to a counter (the initial value of the counter is 0).

The `ACTION` elements `{type:'ADD'}` and `{type:'SUBTRACT'}` correspond to the add button and the subtract button, respectively.

The reducer for this Web page consists of a `switch` statement that contains three `case` statements that handle `ADD`, `SUBTRACT`, and a default operation, as shown here:

```
// two operations: add or subtract
const CountReducer = (state=0, action) => {
   switch(action.type) {
```

```
        case 'ADD':        return state + 1;
        case 'SUBTRACT': return state - 1;
        default:           return state;
    }
};
```

NOTE *Reducers handle state changes synchronously.*

A generalized version of the preceding example involves an `ACTION` element that contains an `amt` field whose numeric value is the amount to add or to subtract from the counter. In this scenario, a typical `ACTION` is `{type:'ADD', amt:3}` or `{pe:'SUBTRACT', amt:5}`, where the value of `amt` is populated elsewhere. Furthermore, the modified version of the `CountReducer` code block is here:

```
const CountReducer = (state=0, action) => {
    switch(action.type) {
        case 'ADD':        return state + action.amt;
        case 'SUBTRACT': return state - action.amt;
        default:           return state;
    }
};
```

A reducer in a real application is similar to the preceding `CountReducer`: The differences involve the specific `case` statements and the code that is executed in each case statement.

A Reducer to Add/Remove Strings from an Array

The following code block defines a reducer that can add an item (`ADD_ITEM`) and remove an item (`DEL_ITEM`) from an array:

```
export const arrReducer = (state = [], action) => {
  switch.action(type) {
     case 'ADD_ITEM': return [...state, action.payload];
     case 'DEL_ITEM':
       return state.filter(n => n.val != action.payload.val);
     default:          return state;
  }
};
```

In the preceding code block a new item is added by creating a new array that contains the current state, with the item appended to that array. Notice how the `spread` operator "..." provides a compact way to list the items in the current state.

An existing item is deleted by conditional logic in a `filter()` method: Everything in the current state, except for the item in question, is returned.

Redux Reducers Are Synchronous

Keep in mind the following points about `Redux` reducers. First, the `arrReducer()` method in the preceding section is a pure function because no mutation occurs in this function; no external variables are required, and a new state is created instead of using the array `push()` method (which is a mutator).

Second, reducers are synchronous; however, you can use redux-observable if you need a reducer that performs calculations asynchronously. Third, data logic in a `reducer()` is separate from the view layer. Fourth, multiple reducers can be defined in an application.

RxFlux is a Flux implementation based on RxJS, and its home page is located here:

https://github.com/fdecampredon/rx-flux

A very good introduction to `Redux` is this video series created by Dan Abramov:

https://egghead.io/series/getting-started-with-redux

The Redux Store

The `Redux Store` maintains the state of an application, which is represented as a global state tree. The `Store` provides two functions: `getState()` and `dispatch()`.

The `getState()` function allows different parts of an application to access the state-related information.

There are two other points to keep in mind regarding the `Store`. First, the store holds the reducers and invokes them in a "broadcast" fashion whenever actions are dispatched. Second, the store provides a subscription mechanism to notify portions of the application (such as the UI) when the state tree has been modified.

The following code block shows you how to create a store:

```
import { createStore } from 'redux'
import RootReducer from '../Reducers/'
const store = createStore(RootReducer)
```

Middleware

The Redux Store can also be extended by plugins in the form of middleware. Various types of third-party middleware are available to perform an assortment of tasks, such as persistence, logging, and flow control.

To summarize, Redux works as follows:

- The application state is encapsulated in a JavaScript object called "state."
- The state is held in a store.
- The store is immutable and never directly changed.
- User interactions fire actions that describe the event and encapsulate the data.

A function called a reducer combines the old state and the action to create the new version of the state, which is kept in the store. Redux simplifies the task of state management by separating the functional code from the presentational code. Instead of using Redux in Angular applications, it's easier to use an implementation of Redux, such as ngrx-store.

https://github.com/mgechev/angular-seed

An extensive description of ngrx-store that also reinforces material from the first part of this chapter can be found here:

https://gist.github.com/btroncone/a6e4347326749f938510

A free 10-minute video about ngrx-store (as this book goes to print) can be found here:

https://egghead.io/lessons/angular-2-ngrx-store-in-10-minutes

This concludes the section regarding Angular applications and Redux. If you decide that Redux is too complex for your needs, one alternative to Redux is MobX, which is simpler than working with Redux. MobX is considered one of the popular (and simpler) alternatives to Redux, and its home page is located here:

https://github.com/mobxjs/mobx

What Is GraphQL?

GraphQL is a data query language and runtime from Facebook that can send data to mobile and Web applications; its home page is located here:

http://graphql.org/

GraphQL is a specification, which means that it can be used with any platform and any language. Facebook maintains its reference implementation, which is written in JavaScript. The `GraphQL` specification is located here:

http://facebook.github.io/graphql/

A `GraphQL` schema acts as a "wrapper" around a data store that can include NoSQL data and relational data. For example, the following link explains how to use `GraphQL` with Mongo:

https://www.compose.io/articles/using-graphql-with-mongodb/

Facebook's StarWars schema is here: *https://goo.gl/oCrK7F*

The following are aspects of GraphQL:

- Query language
- Query semantics
- Query variables
- Mutations
- Fragments

The GraphQL query language is a major part of GraphQL, examples of which you will see in a subsequent section. GraphQL query variables enable you to pass values to GraphQL queries, which is obviously better than using hard-coded values. Mutations allow you to change the dataset behind GraphQL. A mutation is very similar to a field in a GraphQL query, but GraphQL assumes a mutation has side effects and changes the dataset behind the schema. GraphQL fragments provide a mechanism for grouping commonly used fields and reusing them. Some of the programming languages that have implemented `GraphQL` are located here:

http://graphql.org/code/

Companies that use `GraphQL` include Coursera, Intuit, Pinterest, and Shopify. In addition, GitHub supports `GraphQL` (starting from 2016), and has released its public `GraphQL` API:

http://githubengineering.com/the-github-graphql-api/

GraphQL versus REST

As you will see in subsequent sections, `GraphQL` enables you to specify fine-grained queries that return only the data that is required by a client application. Keep in mind the following point about applications that involve GraphQL: Data fetching details are made on the client, whereas data fetching details are made on the server in applications that use Representational State Transfer (REST).

There are several advantages to this approach:

- No redundant data is sent to the client.
- Adding new data fields on the server does not affect queries.
- Only one network round trip is required.

The preceding advantages of GraphQL are particularly important for mobile applications, where the cost of retrieving data can be significant. In addition, the GraphQL query is unaffected by the addition of new fields to the customer object (on the server).

On the other hand, a REST-based request is more coarse-grained and involves "overfetching": Such a request returns 100% of the fields in a customer object, in which no distinction is made between fields that are required and fields that are not required. Moreover, the addition of new fields to a customer object increases the size of the payload returned to the client.

GraphQL Queries

A GraphQL query is a string interpreted by a server that returns data in a specified format. Here is an example of a very simple GraphQL query:

```
{
  emp {
    fname
  }
}
```

The preceding query can be read as "give me the fname attribute of the emp entity." As you can see, the emp entity is followed by a pair of curly braces that contain a single attribute called fname.

The next query is considerably more complex, yet follows the same mechanism as the previous query:

```
{
  user(id: 30000) {
    id,
    name,
    isViewerFriend,
    profilePicture(size: 50) {
      uri,
```

```
      width,
      height
    }
  }
}
```

The preceding query requests the attributes `id`, `name`, and `isViewer-Friend` of the user whose `id` is `30000`. In addition, the query requests the attributes `uri`, `width`, and `height` of the `profilePicture` (50 × 50 pixel size) of the same user.

The response to the preceding query is shown here:

```
{
  "user" : {
    "id": 30000,
    "name": "John Smith",
    "isViewerFriend": true,
    "profilePicture": {
      "uri": "http://www.acme.com/johnsmith.jpg",
      "width": 50,
      "height": 50
    }
  }
}
```

`GraphQL` queries can be very complex, and sometimes the data that is returned contains duplicate "subtrees" of data. The open source project Apollo (discussed later in this chapter) removes the duplicate subtrees from the data that is returned by the server.

Defining a Type System in GraphQL

This section contains an example of a type system that models the JavaScript Object Notation (JSON)-based data in the file `employees.json`.

The first part of the type system defines an `Employee` interface, and the second part of the type system defines a `Query` type, as shown here:

```
interface Employee {
  id: String!
  fname: String
  lname: String
}

type Query {
  emp: Employee
```

```
}
```

For this section, let's make the initial simplifying assumption that there is only one employee named "John Smith." The following query specifies the `fname` field of an employee:

```
{
  emp {
    fname
  }
}
```

The result of the preceding query is shown here:

```
{
  "data": {
    "emp": {
      "fname": "John"
    }
  }
}
```

The following query is a more verbose way of specifying the `fname` field of an employee:

```
query EmpNameQuery {
  emp {
    fname
  }
}
```

The result of the preceding query is shown here:

```
{
  "data": {
    "emp": {
      "fname": "John"
    }
  }
}
```

A query that contains the query keyword and an operation name (such as `EmpNameQuery`) is required to specify multiple fields. For example, the following query specifies the `fname` and the `lname` fields:

```
query EmpNameQuery {
  emp {
    fname
    lname
```

```
    }
  }
```

The result of the preceding query is shown here:

```
{
  "data": {
    "emp": {
      "fname": "John",
      "lname": "Smith"
    }
  }
}
```

Now let's suppose that our data set consists of the following employees:

```
[
  {"fname":"Jane","lname":"Jones","city":"San Francisco"},
  {"fname":"John","lname":"Smith","city":"New York"},
  {"fname":"Dave","lname":"Stone","city":"Seattle"},
]
```

Let's look at the earlier query again:

```
query EmpNameQuery {
  emp {
    fname
    lname
  }
}
```

Now the result of the preceding query involves the `fname` and `lname` fields of three employees, as shown here:

```
{
  "data": {
    {
      "fname": "Jane",
      "lname": "Jones"
    },
    {
      "fname": "John",
      "lname": "Smith"
    },
    {
      "fname": "Dave",
      "lname": "Stone"
    }
  }
}
```

The preceding code samples provide a very basic introduction to the types of queries that you can define in GraphQL. You can also define parameterized queries as well as queries that return a hierarchical dataset.

For example, the following parameterized query specifies the fname, lname, and city fields for the employee whose first name is Jane:

```
{
  Employee(fname: "Jane") {
    fname
    lname
    city
  }
}
```

The preceding query returns the following data:

```
{
  "data": {
    "emp": {
      "fname": "Jane",
      "lname": "Jones"
      "city":  "San Francisco"
    }
  }
}
```

Useful GraphQL Links

Reindex is a GraphQL generator and its home page is located here:

https://www.reindex.io/

Scaphold.io is an online service for creating GraphQL schemas and its home page is located here:

www.scaphold.io

A list of tools and integrations with various languages and data storage engines for GraphQL is located here:

https://www.npmjs.com/search?q=graphql

If you work with React applications, you can use GraphiQL, which is an in-browser integrated development environment (IDE) that is downloadable here:

https://github.com/graphql/graphiql

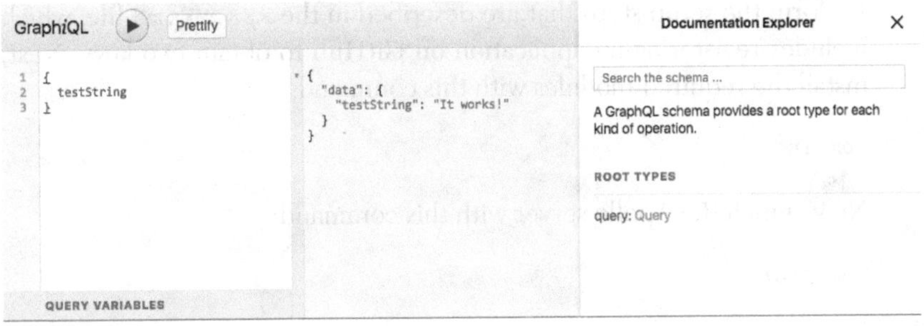

FIGURE 7.2 GraphiQL in a Chrome rrowser.

The following link contains documentation and a code sample for GraphQL:

http://graphql.org/graphql-js/basic-types/

Figure 7.2 displays an example of GraphiQL in a Chrome browser.

The next section discusses Apollo, followed by an application that combines Apollo and Angular.

What Is Apollo?

`Apollo` is an open source project that is designed to work with `GraphQL`, and its home page is located here:

http://www.apollodata.com/

An interactive tutorial (with a video) for Apollo is located here:

https://www.learnapollo.com/introduction/get-started/

The code sample in this section consists of a server component and an Angular-based client component, where the former requires some configuration steps.

NOTE *You must launch the Angular client after you launch the Apollo server, as described in the next two subsections.*

Launching the Apollo Server

Download and uncompress the code from this GitHub repository:

https://github.com/apollographql/GitHunt-API

Perform the setup steps that are described in the `README.md` file, which includes registering an application on GitHub to obtain two keys. Next, install the required modules with this command:

```
npm install
```

Now launch the Apollo server with this command:

```
npm start
```

Figure 7.3 displays the contents that the Apollo server renders (if everything was configured correctly) in a Chrome browser.

Launching the Angular Client

Download and uncompress the Angular client-side code from this GitHub repository:

https://github.com/apollographql/githunt-angular

Next, install the required modules with this command:

```
npm install
```

Now launch the Angular client with this command:

```
npm start
```

Figure 7.4 displays a portion of the contents that the Angular client renders (if everything was configured correctly) in a Chrome browser.

Now let's look at some of the client-side Angular code.

GitHunt API server

Thanks for downloading and running our example server app! This server doesn't include any UI code.

Try one of the following options:

- Go to /graphiql to run some GraphQL queries against this server using GraphiQL.
- Download apollographql/GitHunt-React to run a React-based UI for this app.
- Download apollographql/GitHunt-Angular to run an Angular-based UI for this app.
- Download aruntk/GitHunt-Polymer to run an Polymer-based UI for this app.

Have any improvements in mind? File an issue or a PR about this app at apollographql/GitHunt-API.

FIGURE 7.3 The Apollo aerver in a Chrome browser.

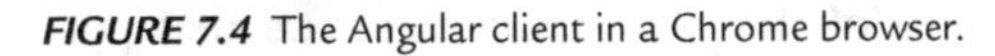

FIGURE 7.4 The Angular client in a Chrome browser.

Project Structure in the Angular Client

The `app/src` subdirectory of the Angular code in the previous section contains the following files:

```
./app.component.html
./app.component.ts
./app.module.ts
./client.ts
./comments
./comments/comment.component.html
./comments/comment.component.ts
./comments/comments-page.component.html
./comments/comments-page.component.ts
./comments/comments-page.model.ts
./comments/index.ts
./feed/feed-entry.component.html
./feed/feed-entry.component.ts
./feed/feed-entry.model.ts
./feed/feed.component.html
./feed/feed.component.ts
./feed/feed.model.ts
./feed/index.ts
./feed/vote-buttons.component.html
./feed/vote-buttons.component.ts
./feed/vote-buttons.model.ts
./index.ts
./navigation/navigation.component.html
./navigation/navigation.component.ts
./new-entry/new-entry.component.html
./new-entry/new-entry.component.ts
```

```
./new-entry/new-entry.model.ts
./profile/profile.component.html
./profile/profile.component.ts
./profile/profile.model.ts
./routes.ts
./shared/index.ts
./shared/info-label.component.ts
./shared/loading.component.ts
./shared/repo-info.component.html
./shared/repo-info.component.ts
./shared/repo-info.model.ts
./subscriptions.ts
```

The file `app.component.ts` contains nothing more than the definition of the `AppComponent` class.

Listing 7.1 displays the contents of the file `app.module.ts` that performs all the work, which differs from the other code samples in this book.

LISTING 7.1: app.module.ts

```
import { NgModule } from '@angular/core';
import { BrowserModule  } from '@angular/platform-browser';
import { FormsModule, ReactiveFormsModule } from '@angular/
                                                     forms';
import { RouterModule } from '@angular/router';
import { ApolloModule }  from 'apollo-angular';
import { EmojifyModule } from 'angular2-emojify';

import { AppComponent }  from './app.component';
import { NavigationComponent }
         from './navigation/navigation.component';
import { ProfileComponent }
         from './profile/profile.component';
import { NewEntryComponent }
         from './new-entry/new-entry.component';
import { FEED_DECLARATIONS } from './feed';
import { COMMENTS_DECLARATIONS } from './comments';
import { SHARED_DECLARATIONS } from './shared';
import { routes } from './routes';
import { provideClient } from './client';
import { InfiniteScrollModule }
         from 'angular2-infinite-scroll';

@NgModule({
  declarations: [
    AppComponent,
    NavigationComponent,
    ProfileComponent,
    NewEntryComponent,
```

```
    ...FEED_DECLARATIONS,
    ...COMMENTS_DECLARATIONS,
    ...SHARED_DECLARATIONS
  ],
  entryComponents: [
    AppComponent
  ],
  imports: [
    BrowserModule,
    FormsModule,
    ReactiveFormsModule,
    RouterModule.forRoot(routes),
    ApolloModule.forRoot(provideClient),
    EmojifyModule,
    InfiniteScrollModule
  ],
  bootstrap: [ AppComponent ],
})
export class AppModule {}
```

Listing 7.1 starts by importing various custom components via the initial import statements, including the Apollo-related `ApolloModule`, which is shown in bold. Next, the `NgModule` decorator contains a `declarations` property that specifies some components, some of which involve the spread operator (which you have not seen in previous examples). The `NgModule` decorator also contains an `imports` property that uses the `ApolloModule.forRoot()` syntax (shown in bold) to reference the exported function `provideClient()` that is defined in `client.ts`. This function handles the details of fetching data from the server.

Various GitHub repositories with examples containing Apollo can be found here:

https://github.com/apollostack

One of the advantages of `Apollo` over `Relay` (discussed later) is its ability to remove duplicate subtrees in the dataset that is returned by a `GraphQL` query. However, Facebook developed Relay, and it's the topic of the next section.

What Is Relay?

Relay is a JavaScript-based technology from Facebook that acts as a "wrapper" around ReactJS components that require data from a server. The Relay home page is located here:

https://github.com/facebook/relay

ReactJS applications can use `Relay` (discussed later in the chapter) to issue data requests to `GraphQL`. In addition, `GraphQL` can provide data to clients (such as Angular clients) that do not use Relay.

NOTE `GraphQL` *can be used independently of* `Relay`, *whereas* `Relay` *cannot be used without* `GraphQL`.

By way of comparison, `REST`-based requests return data for an entity (such as the data about a customer), whereas `Relay` enables you to request individual fields of an entity (such as the first name and last name of a customer). In simplified terms, you can view `Relay` as a finer-grained alternative to `REST`.

Relay uses a network layer to communicate with a GraphQL server. Relay provides a network layer that is compatible with `express-graphql`, and additional features will be added to the network layer as they are developed.

The following link contains an interactive tutorial for Relay:

https://www.learnrelay.org/

NOTE *GraphQL can be used independently of Relay, whereas Relay cannot be used without GraphQL.*

Relay Modern

Facebook released React Fiber in early 2017, which is an extensive rewrite of ReactJS, with the goals of improved performance and extensibility. Facebook also rewrote Relay, which is called Relay Modern. The new features of Relay Modern include static queries, AOT (ahead-of-time) compilation, and built-in garbage collection.

Relay Modern also provides a compatibility API in the event that you are already using an older version of Relay.

Summary

This chapter started with `Flux`, which is a unidirectional pattern for Web applications. You also learned about `Redux`, which is one of the implementations of the `Flux` pattern. In addition, you saw how to use `Flux/Redux` in Angular applications. Next you learned about `Relay` and `GraphQL`, both of which were developed by Facebook. Then you saw an example of an application that uses `GraphQL` to retrieve server-side data that was defined in a `JSON`-based file.

CHAPTER 

ANGULAR AND MOBILE APPS

This chapter explores mobile application development with several toolkits, including Ionic 2, Ionic 2 with NativeScript, Angular with Native Script, and Angular with React Native.

The first part of this chapter discusses Ionic 2. Ionic 2 extends Apache Cordova, which is the open source counterpart to PhoneGap from Adobe. Because Ionic 2 depends on Cordova, you can develop cross-platform hybrid mobile applications. In addition, Ionic 2 uses Angular for the user interface (UI) layer, which is the rationale for its inclusion in this chapter. You will learn how to create a simple Ionic 2 mobile application that displays a list of users. Note that Ionic 2 is the only toolkit in this chapter that relies on a WebView; all the other technologies in this chapter generate cross-platform native mobile applications (i.e., there is no Document Object Model [DOM]).

The second part of this chapter introduces you to `NativeScript`, which you can combine with Ionic 2 to create native mobile applications. This toolkit generates native applications that do not involve a DOM, which is similar to React Native from Facebook.

The third section combines `NativeScript` with Angular to create cross-platform native mobile applications.

The fourth section might surprise you: it provides an overview of React Native for developing cross-platform native mobile applications. Although React Native was initially created in conjunction with ReactJS for the UI layer, you can combine React Native with Angular (for the UI layer) to create cross-platform native mobile applications.

The fifth part of this chapter discusses the Angular Mobile Toolkit, which is based on the Angular CLI, so you can use the `ng` command-line tool to create applications. This toolkit generates mobile applications that are based on Progressive Web Apps (PWAs), which are not discussed in this chapter.

Mobile Development with Ionic 2

Angular is designed to support multiplatform applications, which includes mobile applications. There are also toolkits based on Angular for developing mobile applications. One well-known toolkit is Ionic 2, and its home page is located here:

https://ionic.io

`Ionic` 2 is built on top of Angular and leverages the Cordova toolkit. The `Ionic` framework provides a UI framework that mimics a native UI for creating hybrid mobile applications.

NOTE *Ionic 3 was released as this book went print, and the following link discusses the major changes and new features:*

http://blog.ionic.io/

Installation and Project Creation

```
Install Ionic 2 with the following command:
[sudo] npm install -g ionic cordova
```

If you already have an older version of `Ionic` installed and you encounter an error in the preceding step, the following commands might help you resolve the error:

```
npm uninstall -g ionic
[sudo] npm i -g ionic cordova
```

Now create an `Ionic` application with the following command:

```
ionic start firstIonicProject blank --v2
```

The name of our project is `firstIonicProject`, and it's based on the blank `Ionic` template. The argument `--v2` indicates `Ionic` 2 (otherwise the default is `Ionic` 1).

```
cd firstIonicProject
```

Start the `Ionic` server with this command:

```
ionic serve
```

You will see the following type of output from the preceding command:

```
> ionic-hello-world@ ionic:serve firstIonicProject
> ionic-app-scripts serve "--v2" "--address" "0.0.0.0"
"--port" "8100" "--livereload-port" "35729"

[13:31:04]  ionic-app-scripts 1.1.4
[13:31:04]  watch started ...
[13:31:04]  build dev started ...
[13:31:04]  clean started ...
[13:31:04]  clean finished in 1 ms
[13:31:04]  copy started ...
[13:31:04]  transpile started ...
[13:31:17]  transpile finished in 13.07 s
[13:31:17]  preprocess started ...
[13:31:17]  preprocess finished in 1 ms
[13:31:17]  webpack started ...
[13:31:18]  copy finished in 14.11 s
[13:31:33]  webpack finished in 15.74 s
[13:31:33]  sass started ...
[13:31:36]  sass finished in 3.17 s
[13:31:36]  postprocess started ...
[13:31:36]  postprocess finished in 1 ms
[13:31:36]  lint started ...
[13:31:36]  build dev finished in 32.05 s
[13:31:36]  watch ready in 32.18 s
[13:31:36]  dev server running: http://localhost:8100/
```

In addition, a browser session is automatically launched at `local-host:8080`, where you will see the contents of Figure 8.1.

You can also view a simulation of the application on a mobile device by launching the following command:

```
ionic serve -l
```

The preceding command launches a browser at `localhost:8100` and displays the output on a simulated iPhone and Android device, where the latter display was selected from the drop-down list in the top-right corner).

Figure 8.2 displays the simulated iPhone and Android devices in a browser.

The next section briefly discusses Ionic native mobile applications.

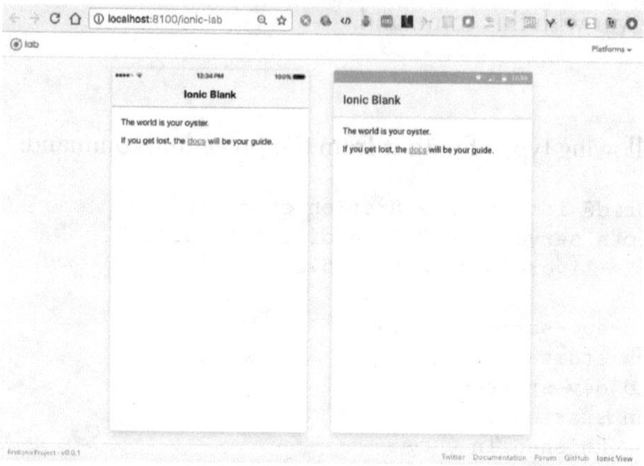

FIGURE 8.1 A minimal Ionic application in a Chrome browser.

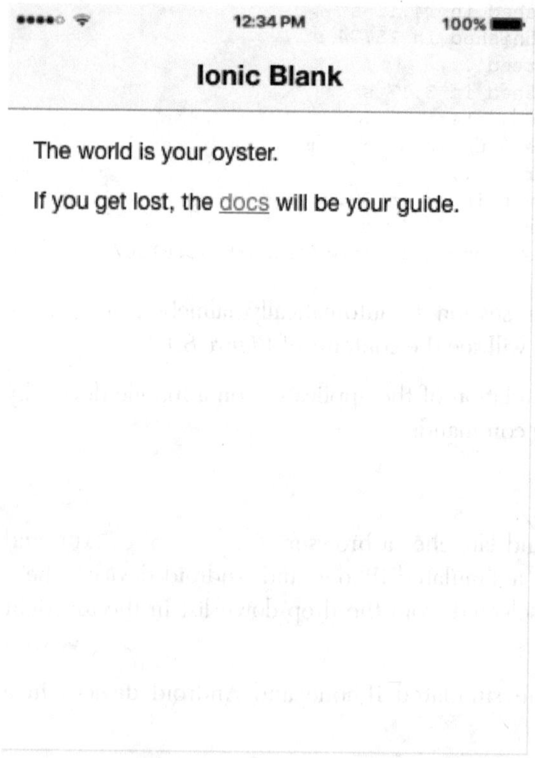

FIGURE 8.2 An Ionic app simulated on iOS and Android in a Chrome browser.

Ionic Native

The Ionic Native home page is located here:

http://ionicframework.com/docs/v2/native/

Ionic Native wraps plugin callbacks in a `Promise` or an `Observable`, providing a common interface for all plugins and ensuring that native events trigger change detection in Angular. The following code block illustrates how to add Geolocation functionality to an Ionic Native application:

```
import {Geolocation} from 'ionic-native';

Geolocation.getCurrentPosition().then(pos => {
  console.log('lat: ' + pos.coords.latitude + ', lon: ' + pos.
                                                coords.longitude);
});

let watch = Geolocation.watchPosition().subscribe(pos => {
  console.log('lat: ' + pos.coords.latitude + ', lon: ' + pos.
                                                coords.longitude);
});

// to stop watching
watch.unsubscribe();
```

The next section discusses the project structure and the contents of the TypeScript file `app.component.ts`.

The Project Structure of Ionic Applications

The directory `firstProject` in the previous section contains the following files and folders:

```
> hooks
> node_modules
> platforms
> plugins
> resources
> src
  > app
    app.component.ts
    app.html
    app.module.ts
    app.scss
    main.ts
  > assets
```

```
    > pages
    > theme
  > www
```

The app/src directory contains familiar TypeScript files that you have seen in code samples in earlier chapters. Listing 8.1 displays the contents of app.component.ts, whose contents are significantly different from Angular Web applications.

LISTING 8.1: app.component.ts

```
import { Component } from '@angular/core';
import { Platform } from 'ionic-angular';
import { StatusBar, Splashscreen } from 'ionic-native';
import { HomePage } from '../pages/home/home';

@Component({
  templateUrl: 'app.html'
})
export class MyApp {
  rootPage = HomePage;

  constructor(platform: Platform) {
    platform.ready().then(() => {
      // Okay, so the platform is ready and our plugins are
available.
      // Here you can do any higher level native things you
might need.
      StatusBar.styleDefault();
      Splashscreen.hide();
    });
  }
}
```

Listing 8.1 starts with a standard import statement, followed by three Ionic-specific import statements. Notice that the @Component decorator does not contain a selector property, which is required in Angular applications. The templateUrl property references the HTML page app.html whose contents are very similar to the Web page index.html in Cordova applications.

Listing 8.1 exports the class MyApp, which initializes the rootPage variable (specific to Ionic), followed by a constructor. If you have worked with Cordova, the ready() method in the constructor probably looks familiar to you. However, in Ionic applications, platform.ready() returns a Promise, and when the latter is resolved, the then() method displays the main screen based on the contents of app.html.

Listing 8.2 displays the contents of `app.module.ts`, which has a familiar structure (e.g., the `@NgModule` decorator), along with Ionic-specific contents.

LISTING 8.2: app.module.ts

```
import { NgModule, ErrorHandler } from '@angular/core';
import { IonicApp, IonicModule, IonicErrorHandler }
         from 'ionic-angular';
import { MyApp } from './app.component';
import { HomePage } from '../pages/home/home';

@NgModule({
  declarations: [
    MyApp,
    HomePage
  ],
  imports: [
    IonicModule.forRoot(MyApp)
  ],
  bootstrap: [IonicApp],
  entryComponents: [
    MyApp,
    HomePage
  ],
  providers: [{provide: ErrorHandler, useClass:
IonicErrorHandler}]
})
export class AppModule {}
```

Listing 8.2 imports the `MyApp` class (shown in Listing 8.1) and the `HomePage` class (not shown here), both of which are listed in the `declarations` property. You can read the Ionic documentation to learn about the other components in Listing 8.2.

The next section shows you how to retrieve GitHub user-related information (similar to the sample in Chapter 4) via an Ionic application.

Retrieving GitHub User Data in an Ionic Application

Copy the `IonicGithub` directory from the companion disc to a convenient location. This application contains the following files that have custom code, all of which are in the `src/app` subdirectory:

github.ts
app.component.ts

app.module.ts
app.html

Listing 8.3 displays the contents of `github.ts` that defines the custom component `GitHubService`, which retrieves GitHub-related information about a given user.

LISTING 8.3: github.ts

```
import {Injectable}    from '@angular/core';
import {Http, Headers} from '@angular/http';

@Injectable()
export class GitHubService {
  constructor(private http: Http) {
  }

  getRepos(username) {
          let repos = this.http.get('https://api.github.
                              com/users/${username}/repos');

      return repos;
  }
}
```

Listing 8.3 defines and exports the custom class `GitHubService`, which contains the method `getRepos()` for retrieving the list of GitHub repositories for a given user. In this application, the user has the hard-coded value ocampesato, which you would replace with an `<input>` element to accept an arbitrary GitHub user name.

The method `getRepos()` invokes the `get()` method of the `http` instance variable that is instantiated as a private variable in the constructor, and then returns the result of that method's invocation.

Listing 8.4 displays the contents of `app.component.ts` that defines and exports the `MyApp` component that invokes the `getRepos()` method in the `GitHubService` class.

LISTING 8.4: app.component.ts

```
import { Component } from '@angular/core';
import { Platform } from 'ionic-angular';
import { StatusBar, Splashscreen } from 'ionic-native';
import { HomePage } from '../pages/home/home';
import {GitHubService} from './github';
```

```
@Component({
  templateUrl: 'app.html'
})
export class MyApp {
  rootPage = HomePage;
  public repoList;
  public username = "ocampesato";


  constructor(platform: Platform, private github:
                                            GitHubService) {
    platform.ready().then(() => {
      // Okay, so the platform is ready and our plugins are
available.
      // Here you can do any higher level native things you
might need.
      StatusBar.styleDefault();
      Splashscreen.hide();
    });
  }

  getRepos() {
    this.github.getRepos(this.username).subscribe(
        data => {
            this.repoList = data.json();
        },
        err => console.error(err),
        () => console.log('getRepos completed')
    );
  }
}
```

Listing 8.4 modifies the default constructor by adding the variable `github`, which is a private instance of the `GitHubService` class (defined in Listing 8.3). The `getRepos()` method is invoked when users click the button in the HTML Web page. This method subscribes to the `Observable` that is returned from the `getRepos()` method of the `GitHubService` class.

Listing 8.5 displays the contents of `app.module.ts` that defines and exports the `AppModule` component.

LISTING 8.5: app.module.ts

```
import { NgModule, ErrorHandler } from '@angular/core';
import { IonicApp, IonicModule, IonicErrorHandler }
         from 'ionic-angular';
import { MyApp }        from './app.component';
```

```
import { HomePage }    from '../pages/home/home';
import {GitHubService} from './github';

@NgModule({
  declarations: [
    MyApp,
    HomePage
  ],
  imports: [
    IonicModule.forRoot(MyApp)
  ],
  bootstrap: [IonicApp],
  entryComponents: [
    MyApp,
    HomePage
  ],
  providers: [{provide: ErrorHandler, useClass:
IonicErrorHandler}, GitHubService]
})
export class AppModule {}
```

Listing 8.5 contains code that is similar to the previous Ionic code sample, with the addition of the `import` statement for `GitHubService` and the modified contents of the `providers` property, both of which are shown in bold.

Listing 8.6 displays the contents of `app.html` that contains the HTML markup to display the list of repositories for a GitHub user.

LISTING 8.6: app.html

```
<ion-navbar *navbar>
    <ion-title>
        GitHub
    </ion-title>
</ion-navbar>

<ion-content class="home">
    <ion-list inset>
        <ion-item>
            <ion-label>Username</ion-label>
            <ion-input [(ngModel)]="username" type="text">
            </ion-input>
        </ion-item>
    </ion-list>

    <div padding>
        <button block (click)="getRepos()">Search</button>
    </div>
```

```
    <ion-card *ngFor="let repo of foundRepos">
        <ion-card-header>
            {{ repo.name }}
        </ion-card-header>
        <ion-card-content>
            {{ repo.description }}
        </ion-card-content>
    </ion-card>
</ion-content>
```

Listing 8.6 contains three sections, the first of which displays the name of the GitHub user that is used to perform a GitHub query. The second section contains a `<div>` element with a nested `<button>` element that invokes the `getRepos()` method when users click this button. The third section contains an Ionic `<ion-card>` element with an `*ngFor` directive that iterates through the list of GitHub repositories and displays each repository in its own card-like container.

Figure 8.3 displays the list of repositories of a GitHub user.

The following link describes the differences between PhoneGap, Ionic, Titanium (not discussed in this chapter), and Cordova:

> *https://maheshkariya.wordpress.com/2017/04/18/what-are-the-basic-differences-between-phonegap-ionic-titanium-and-cordova*

The next option for developing cross-platform native mobile applications is the combination of Ionic 2 and NativeScript. Before looking at

Username ocampesato

Search

android-beginner
Android for Beginners

android-beginners

android-graphics
android graphics

angular-graphics
AngularJS and SVG graphics

angular1x-beginners
Angular 1.x for Beginners

FIGURE 8.3 The list of repositories of a GitHub user.

a code sample, let's take a quick detour to learn some basic features of NativeScript, which is the topic of the next section.

What Is NativeScript?

NativeScript 2.0 enables you to create native mobile applications for multiple platforms, and its home page is located here:

https://www.nativescript.org/

NativeScript renders UIs with the native platform's rendering engine (i.e., no WebViews), which results in native-like performance. However, NativeScript generates native applications for Android and iOS, whereas Ionic 2 generates hybrid applications for Android and iOS. Hence, NativeScript is "comparable" to React Native, whereas Ionic 2 is "comparable" to PhoneGap.

`NativeScript` has the following features:

- Native UI (no WebViews)
- Extensible
- Quick to learn
- Cross-platform
- Backed by Telerik
- Open source (Apache 2 license)

Keep in mind that `NativeScript` for Angular generates native applications for Android and iOS, whereas Ionic 2 generates hybrid applications for Android and iOS. Hence, the combination of `NativeScript` and Angular is a "counterpart" to React Native (where the UI layer can be either ReactJS or Angular).

Installation and New Project Creation

Install `NativeScript` with this command:

```
[sudo] npm install -g nativescript
```

After the preceding command has completed, you can create and deploy a NativeScript application by executing the following three commands:

```
tns create FirstNSProject --template
nativescript-template-tutorial
cd FirstNSProject
tns run android
```

For simplicity, the `FirstNSProject` project is based on an existing NativeScript template.

Invoke the following command to deploy to an iOS device:

```
tns run ios
```

Figure 8.4 displays the (admittedly trivial) contents of the NativeScript `FirstNSProject` on an Android device.

A list of NativeScript plugins is located here:

http://plugins.telerik.com/nativescript

In case you're wondering, the text in Figure 8.4 is specified in the file `main-page.xml` (in the `app` subdirectory), whose contents are shown here:

```
<Page loaded="pageLoaded">
  <ActionBar title="My App" class="action-bar"></ActionBar>
  <!-- Your UI components go here -->
</Page>
```

As you can see, NativeScript uses an XML file to define the contents and layout of NativeScript applications.

My App

FIGURE 8.4 A NativeScript application on an Android device.

You can also convert an existing Ionic 2 application to NativeScript:

https://www.thepolyglotdeveloper.com/2016/05/converting-ionic-2-mobile-app-nativescript

Angular and NativeScript

Install `NativeScript` as described in the previous section, in case you have not already done so, and then perform the following steps:

```
tns create FirstNSNGProject --template
nativescript-template-ng-tutorial
cd FirstNSNGProject
tns run android
```

Invoke the following command to deploy a `NativeScript`-based iOS application to an iOS device:

```
tns run ios
```

If you prefer to launch applications in a simulator, invoke one of the following commands:

```
tns run android --emulator
```

This concludes the section of the chapter regarding NativeScript. The next section delves into React Native, which is a cross-platform toolkit for generating native mobile applications for Android and iOS.

Working with React Native

React Native is a toolkit developed by Facebook for developing mobile applications, and its home page is located here:

https://facebook.github.io/react-native/

Because this book is about Angular, why does this chapter contain React Native? The answer might surprise you: It's possible to create native mobile applications using a combination of Angular and React Native, which you will see later in this chapter. Nevertheless, this section might be optional for some readers, and feel free to skip this section.

Another point to keep in mind: React Native development requires at least a basic understanding of ReactJS. In some cases, it's useful to know how to deploy a mobile application to a device from Xcode and from Android Studio.

Setup Steps for React Native

Before you can create and deploy React Native applications to iOS devices and Android devices, make sure that you have the following hardware and software installed (other versions are supported in Step 1):

1) a MacBook with El Capitan (10.11.4), and Xcode 7.3
2) an iOS device with El Capitan
3) a MacBook with Android Studio 2.x

Next, perform the following setup steps:

4) install the Android SDK, Android NDK, and Java8 on your MacBook
5) set the environment variables `ANDROID_HOME`, `NDK_HOME`, and `JAVA_HOME`

You also need Android 4.x or higher installed on an Android device. Next, install `node` on your MacBook if you haven't done so already, and then install React Native on your MacBook with this command:

```
npm install -g react-native-cli
```

The following links contain more detailed setup instructions for React Native:

1) *https://facebook.github.io/react-native/docs/getting-started.html#-content*
2) *https://facebook.github.io/react-native/docs/android-setup.html*
3) *https://facebook.github.io/react-native/docs/running-on-device-android.html*

The next section shows you how to create and deploy a React Native application to a mobile device.

How to Create a React Native Application

After you have completed the setup steps in the previous sections, create a new React Native project as follows:

1) Navigate to a convenient location.
2) Invoke this command: `react-native FirstRNProject.`
3) Navigate into the root directory: `cd FirstRNProject`

Now install the dependencies with this command (which will take a while):

```
npm install
```

After the preceding command has completed, you can view the contents of `FirstRNProject`, which are shown below for your convenience:

```
android/
index.android.js
index.ios.js
ios/
node_modules/
package.json
```

Listing 8.7 displays the contents of `index.android.js`, which is similar to the contents of `index.ios.js`.

LISTING 8.7: index.android.ts

```
/**
 * Sample React Native App
    * https://github.com/facebook/react-native
 */

import React, { Component } from 'react';
import {
  AppRegistry,
  StyleSheet,
  Text,
  View
} from 'react-native';

export default class MyRNProject extends Component {
  render() {
    return (
      <View style={styles.container}>
        <Text style={styles.welcome}>
          Welcome to React Native!
        </Text>
        <Text style={styles.instructions}>
          To get started, edit index.android.js
        </Text>
        <Text style={styles.instructions}>
          Double tap R on your keyboard to reload,{'\n'}
          Shake or press menu button for dev menu
        </Text>
      </View>
    );
  }
```

```
}

const styles = StyleSheet.create({
  container: {
    flex: 1,
    justifyContent: 'center',
    alignItems: 'center',
    backgroundColor: '#F5FCFF',
  },
  welcome: {
    fontSize: 20,
    textAlign: 'center',
    margin: 10,
  },
  instructions: {
    textAlign: 'center',
    color: '#333333',
    marginBottom: 5,
  },
});

AppRegistry.registerComponent('MyRNProject', () =>
MyRNProject);
```

Listing 8.7 starts with `import` statements for various required components, including the `StyleSheet` component that is needed later in the code. The next section of Listing 8.7 defines the `MyRNProject` component that contains a `render()` method, which returns a top-level `<View>` component. Notice that the `<View>` component contains three `<Text>` components that contain plain text strings. The third section of Listing 8.5 defines the variable `styles` from the `StyleSheet` component, which in turn defines three properties that are referenced in the three `<Text>` components.

The final code snippet in Listing 8.7 invokes the `registerComponent()` method of the `AppRegistry` component in order to register the custom `MyRNProject` class: This step is required for React Native applications but not for ReactJS.

Deploying React Native Apps to a Mobile Device

Deploy the native Android application from the previous section to an Android device as follows:

```
1) adb devices (check that a device is available)
2) cd <top-of-project>
3) react-native run-android
```

Deploy the native iOS application from the previous section to an iOS device as follows:

```
1) cd <top-of-project>
2) react-native run-ios
```

You can refresh the contents of the screen with this command:

```
command-r
```

Now that you have completed this introductory section, you are ready to see how to create a mobile application that uses Angular and React Native, which is the topic of the next section.

Angular and React Native

A GitHub repository with code that combines Angular with React Native can be found here:

https://github.com/angular/react-native-renderer

Download the code from the preceding link and perform the following steps (which are listed in the `README.md` file):

```
npm install -g gulp react-native-cli typings
npm install
gulp init
gulp start.android (for android)
gulp start.ios (for ios)
```

NOTE *Make sure that you set* `JAVA_HOME` *for Java8 (not Java9) in the command shell where you launch the preceding list of commands.*

Listing 8.8 displays the contents of `NG2RNGraphics.ts` that illustrates how to generate a set of colored circles that follow the path of an Archimedean spiral. Note that the code contains portions of `animate.js` in the `sample/samples/android` subdirectory of the GitHub repository.

LISTING 8.8: NG2RNGraphics.ts

```
import {Component, Input, Output, ElementRef, EventEmitter,
ViewChildren, QueryList} from '@angular/core';
import {StyleSheet} from 'react-native';

@Component({
```

```
  selector: 'ball',
  inputs: ['color', 'x', 'y', 'radius'],
  template: `
<View [styleSheet]="styles.ball" [style]="{top: _y, left:
_x, backgroundColor: _color, borderRadius: _radius, width: _
                                    radius*2, height: _radius*2}"
  (pan)="moveBall($event)" (panend)="endMoveBall($event)">
</View>
`
})
export class Ball {
  _x: number;
  _y: number;
  _color: number;
  _radius: number;
  _vX: number = 0;
  _vY: number = 0;
  styles: any;
  _el: any;
  constructor(el: ElementRef) {
    this._el = el.nativeElement;
    this.styles = StyleSheet.create({
      ball: {
        position: 'absolute'
      }
    });
  }

  set x (value: any) { this._x = (!isNaN(parseInt(value))) ?
                                   parseInt(value) : value;  }

  set y (value: any) { this._y = (!isNaN(parseInt(value))) ?
                                   parseInt(value) : value; }

  set radius (value: any) { this._radius =
       (!isNaN(parseInt(value))) ? parseInt(value) : value; }

  set color (value: any) { this._color = value;}
}

@Component({
  selector: 'animation-app',
  host: {position: 'absolute', top: '0', left: '0', bottom:
                                        '0', right: '0'},
  template: '
    <ball *ngFor="let ball of balls" x="{{ball.x}}"
y="{{ball.y}}" color="{{ball.color}}" radius="{{ball.
                                        radius}}"</ball>
  '
})
export class AnimationApp {
```

```
  @ViewChildren(Ball) ballsChildren: QueryList<Ball>;
  balls: Array<any> = [];
  basePointX  = 200;
  basePointY  = 150;
  currentX    = 0;
  currentY    = 0;
  offsetX     = 0;
  offsetY     = 0;
  radius      = 0;
  smallRadius = 20;
  lineWidth   = 2;
  spiralCount = 4;
  angle       = 0;
  Constant    = 0.25;
  deltaAngle  = 2;
  maxAngle    = 721;
  rectWidth   = 40;
  rectHeight  = 20;
  index       = 0;
  ballColors  = ['#f00', '#ff0'];

  constructor(){
   for(this.angle=0; this.angle<this.maxAngle;
                     this.angle+=this.deltaAngle) {
     this.radius   = this.Constant*this.angle;
     this.offsetX  = this.radius*Math.cos(this.angle*Math.
                                                  PI/180);
     this.offsetY  = this.radius*Math.sin(this.angle*Math.
                                                  PI/180);
     this.currentX = this.basePointX+this.offsetX;
     this.currentY = this.basePointY-this.offsetY;

     this.index = Math.floor(this.angle/this.deltaAngle);

     this.balls.push({
        x:      this.currentX,
        y:      this.currentY,
        radius: this.smallRadius,
        color:  this.ballColors[this.index%this.ballColors.
                                                   length]
     });
   }
  }
}
```

Listing 8.8 starts by importing Angular modules as well as a `Stylesheet` (for cascading style sheet [CSS]-related properties) from `react-native`. Next, the `Component` decorator for the `Ball` class contains a template property that consists of a `View` component that specifies various attributes, such as the stylesheet, CSS properties, along with the methods `moveBall()` and `endMoveBall()` to handle the pan event and end-of-pan events, respectively. The `constructor()` of the `Ball` class performs some initialization, and the set methods ensure that the `x, y, radius,` and `color` properties have valid values.

The second part of Listing 8.8 contains the `Component` decorator for the `AnimationApp` class that initializes the `selector` property with a custom element and the `host` property with CSS-related values. In addition, the template property contains an `*ngFor` statement that iterates through the collection of `Ball` instances that create the graphics effect.

The `AnimationApp` class initializes multiple TypeScript variables, followed by a `constructor()` that contains a loop that populates the TypeScript array `balls` with circles that follow the path of an Archimedean spiral.

You can make a backup of `animation.js` in the `sample/samples/android` subdirectory and replace its contents with `NG2RNGraphics.ts`, or you can clone the directory and work with a separate project.

Additional documentation for this repository is located here:

http://angular.github.io/react-native-renderer/

Figure 8.5 displays the output from launching the code in Listing 8.8 on an Android device.

React Native versus NativeScript: A High-Level Comparison

Although both of these toolkits enable you to develop native mobile applications, they do have differences. First, NativeScript combines the higher-level functionality of Angular with the lower-level UI elements in NativeScript to create cross-platform applications that have a consistent look and feel to them. On the other hand, React Native enables developers to write platform-agnostic code and simultaneously access the platform-specific UI layer. Second, React attempts to provide an abstraction for business logic, whereas NativeScript provides a more "unified" development experience.

Third, React Native provides rapid deployment and execution via "hot reloading," which (in some cases) might be superior to NativeScript.

FIGURE 8.5 Graphics from an Angular and React Native app on an Android device.

Fourth, the use of Angular with NativeScript requires "embracing" an Angular architecture, which is not the case for React. NativeScript and Angular are also separate open source projects, and hence another dependency, whereas React Native handles cross-platform functionality in the React framework.

These points will assist you in weighing the various trade-offs, in conjunction with your project requirements, to determine the environment that best suits your needs.

Angular Mobile Toolkit

The Angular Mobile Toolkit (AMT) enables you to create Progressive Web Apps, and its home page is located here:

https://mobile.angular.io/

The GitHub link with downloadable code is located here:

https://github.com/angular/mobile-toolkit

The following quote (from the README.md file) is a succinct description of this GitHub repository:

> *This repo is a series of tools and guides to help build Progressive Web Apps. All guides are currently based on Angular CLI, and all tools should be considered alpha quality. In the future, more guides and recipes to cover different tools and use cases will be added here and on [https://mobile.angular.io].*

If you are unfamiliar with Progressive Web Apps (PWAs are discussed very briefly later in this chapter), a good introduction is located here:

> *https://developers.google.com/web/fundamentals/getting-started/codelabs/your-first-pwapp/#download_the_code*

The next several sections cover the following topics:

- Creating an AMT project with `ng`
- Contents of the `manifest.webapp` file
- Installing the mobile application
- Building the app shell
- Adding offline capabilities via Service Workers

Creating a Mobile Project

Complete the following steps to create a PWA-based mobile application in Angular via the `ng` command-line utility:

Step 1: Install the Angular CLI as described in Chapter 1.

Step 2: Create a new project with this command:

```
ng new hello-mobile --mobile
```

Step 3: Serve the mobile application:

```
cd hello-mobile
ng serve
```

Step 4: Navigate to `localhost:4200` (usually automatic).

When you invoke `ng` with the `--mobile` flag, the `ng` utility creates a PWA with the following:

- A Web Application Manifest with information for installing your application on the home screen
- A build step to generate an App Shell from the root component

- A Service Worker script to automatically cache your application for fast loading (even without an Internet connection)

Keep in mind that the Service Worker is only installed in production mode, which means either `ng serve --prod` or `ng build --prod`.

The manifest.webapp File

The code sample in the previous section contains a Web App Manifest that is automatically generated during project creation. Listing 8.9 displays the contents of the `manifest.webapp` file in the `src` subdirectory.

LISTING 8.9: manifest.webapp

```
{
  "name": "Hello Mobile",
  "short_name": "Hello Mobile",
  "icons": [
    {
          "src": "/android-chrome-36x36.png",
          "sizes": "36x36",
          "type": "image/png",
          "density": 0.75
      },
      {
          "src": "/android-chrome-48x48.png",
          "sizes": "48x48",
          "type": "image/png",
          "density": 1
      },
      {
          "src": "/android-chrome-72x72.png",
          "sizes": "72x72",
          "type": "image/png",
          "density": 1.5
      },
      {
          "src": "/android-chrome-96x96.png",
          "sizes": "96x96",
          "type": "image/png",
          "density": 2
      },
      {
          "src": "/android-chrome-144x144.png",
          "sizes": "144x144",
          "type": "image/png",
          "density": 3
```

```
        },
        {
            "src": "/android-chrome-192x192.png",
            "sizes": "192x192",
            "type": "image/png",
            "density": 4
        }
    ],
    "theme_color": "#000000",
    "background_color": "#e0e0e0",
    "start_url": "/index.html",
    "display": "standalone",
    "orientation": "portrait"
}
```

Listing 8.9 contains JavaScript Object Notation (JSON)-based data, the longest of which is the `icons` property, which is an array consisting of the properties of six PNG files. Listing 8.9 also contains color-related properties, as well as the `start_url` property (which specifies the HTML Web page `index.html`).

Navigate to the following GitHub link for information about installing the mobile app, building the app shell, and adding offline capabilities:

https://github.com/angular/mobile-toolkit

Progressive Web Apps (Optional)

Recently, PWAs have gained significant interest in the mobile community. If you are unfamiliar with PWAs, here is a quote from Wikipedia:

> *Progressive Web App (PWA) is a term used to denote a new software development methodology. Unlike traditional applications, Progressive Web App can be seen as an evolving hybrid of regular web pages (or websites) and a mobile application. This new application life-cycle model combines features offered by most modern browsers with benefits of mobile experience. (https://en.wikipedia.org/wiki/Progressive_web_app)*

PWAs are mobile-like native applications that are created from HTML5, CSS3, JavaScript, and Service Workers. PWAs load quickly, can work offline, and support push notifications. An icon for a PWA appears on mobile devices, just like a "regular" native mobile application. However, PWAs are not discoverable in any application store, and depending on your perspective, this can be an advantage (e.g., the publication process is obviated) or a disadvantage (e.g., other people cannot find your PWAs, which might reduce traffic).

Currently, Chrome provides support for PWAs, and potentially other browsers will also support PWAs at some point in the future.

Web Workers and Service Workers

A Service Worker provides the functionality of a Web Worker, along with additional features. You can learn about Service Workers here:

https://developers.google.com/web/fundamentals/getting-started/primers/service-workers?hl=en

Components of a PWA

A PWA often consists of the following files:

- an index.html Web page,
- an app.js file for navigation and UI logic,
- an application shell file, and
- a cache file.

You can search online for blog posts that display the contents of some of the files in the preceding list. In addition, you can peruse an assortment of PWAs at this site:

https://pwa.rocks/

Other Links

The following link contains a comparison of PWAs and Android:

http://stackoverflow.com/questions/35504194/what-features-do-progressive-web-apps-have-vs-native-apps-and-vice-versa-on-an

The following link contains information about Angular and PWAs:

http://www.slideshare.net/ManfredSteyer/progressive-web-apps-with-angular-2

Summary

This chapter showed you a multitude of ways to create mobile applications that involve Angular for the UI layer. First you learned about the

Angular Mobile Toolkit, which enables you to create mobile applications with Angular. You also learned how to use the Ionic 2 toolkit to create cross-platform hybrid mobile applications. Then you saw how to combine Ionic 2 with NativeScript to create native mobile applications.

You also saw how to create native mobile applications with Angular and NativeScript. Next, you learned about React Native, which enables you to generate cross-platform native mobile applications with either ReactJS or Angular for the UI layer.

CHAPTER 9

FUNCTIONAL REACTIVE PROGRAMMING

This chapter discusses functional reactive programming (FRP), with a focus on basic `RxJS` code samples that can help you develop simple Web applications (i.e., even without Angular). Starting from Chapter 4, you have seen various Angular code samples that use `Observables`. However, this chapter delves into other aspects of `RxJS` that were not covered in previous chapters.

As you will soon see, most of the code samples in this chapter are simple HTML Web pages. This approach significantly reduces development time because you simply launch the HTML Web pages in a browser without having to create applications from the command line (and also there is no need for the `node_modules` subdirectory). Thus, this chapter enables you to focus on quickly learning various aspects of RxJS, after which you can use the features that you need in your Angular applications.

The first part of the chapter contains a high-level introduction to FRP, along with a list of some popular JavaScript toolkits for FRP.

The second part of this chapter discusses intermediate operators and terminal operators in `RxJS`, including code samples that illustrate how to invoke multiple operators via method chaining. This section also discusses the difference between a cold `Observable` and a hot `Observable`, as well as how you can't "convert" the former into the latter. Almost all the code samples in this section are complete and self-contained, so you can launch them in a browser and view their output in Chrome Web Inspector.

The third part of this chapter contains examples of using `RxJS` with scalable vector graphics (SVG) to generate graphics and animation effects. You will also see an example of "reactifying" some HTML elements in an HTML Web page.

The final part of this chapter provides a very brief section regarding version 5 of `RxJS`, along with some differences between version 5 and version 4 of `RxJS`.

What Is Functional Reactive Programming (FRP)?

Various definitions of FRP are available on the Web. For our purposes, FRP is based on a combination of the `Observer` pattern, the `Iterator` pattern, and functional programming. The home page is located here:

http://reactivex.io/

Reactive programming was introduced in 1997, and can be summarized as follows:

- It is programming with asynchronous data streams.
- It is event-driven instead of proactive.
- Multiple toolkits and libraries are available.
- It supports languages such as JS, Java, Scala, Android, and Swift.

Conal Elliott is credited with creating FRP, and you can find his very specific definition of FRP here:

https://stackoverflow.com/questions/1028250/what-is-functional-reactive-programming

Another definition of FRP involves a combination of two other concepts: reactive programming (asynchronous data streams) and functional programming (pure functions, immutability, and minimal use of variables and state).

Reactive programming supports a number of operators that provide powerful functionality when working with asynchronous streams. The reactive programming paradigm avoids the "callback hell" that can occur in other environments. Moreover, `Observables` provide greater flexibility than working with `Promise`-based toolkits and libraries.

FRP is partially based on functional programming, which has gained popularity because it can reduce the amount of state in a program, which can in turn help reduce the number of code bugs. However, FRP is more declarative than functional programming, usually has more abstraction, and can involve higher-order functions. Consequently, the combination of reactive programming and functional programming enables you to write more succinct yet powerful code.

According to the Reactive home page, FRP handles errors properly in asynchronous streams and avoids the necessity of writing custom code to deal with threads, synchronization, and concurrency. From another perspective, FRP is the "culmination" of the path from `Collections`, then to `Streams`, and finally to asynchronous `Streams`.

Several toolkits for FRP in JavaScript can be found at the following sites:

RxJS: *https://github.com/Reactive-Extensions/RxJS*

Bacon.js: *https://baconjs.github.io/*

Kefir.js: *https://rpominov.github.io/kefir/*

most.js: *https://github.com/cujojs/most*

The preceding toolkits have different strengths and are typically more lightweight than RxJS. After you have completed this chapter, you will be in a better position to evaluate these alternatives to RxJS, and whether you want to use them instead of RxJS.

Now let's take a brief look at the `Observer` pattern that is fundamental to FRP.

The Observer Pattern

The `Observer` pattern is a powerful pattern that is implemented in many programming languages. In simplified terms, the `Observer` pattern involves an `Observable` (i.e., something that is observed or "watched") and one or more `Observer` objects. An `Observer` (also called a subscriber) "watches" for changes in data or the occurrence of events in another object. In languages such as Java, an `Observable` contains a collection of `Observer` objects that have registered themselves with the `Observable`. When a state change or an event occurs in the `Observable`, the `Observable` notifies the registered `Observer` objects.

The details of defining `Observables` are discussed later in this chapter, but the key idea involves combining ("chaining") operators (such as `map()` and `filter()`) and then invoking the `subscribe()` method to "make it happen."

Handling Asynchronous Events

RxJS is sometimes described as "LoDash for asynchronous events." Various types of asynchronous events include the following:

- Ajax
- User events (including mouse-related events)
- Animation
- Sockets and server-sent events (SSEs)
- Workers

By way of comparison, the following code snippet illustrates the ECMA5 style for handling an asynchronous event:

```
getDataFromSomewhere(function(result) {
    console.log("result = "+result)
});
```

The equivalent ES6 version of the preceding code snippet is shown here:

```
getDataFromSomewhere((result) => {
    console.log("result = "+result)
});
```

Promises and Asynchronous Events

`Promises` are well-suited for asynchronous operations (such as Ajax-based requests), provided that the expected behavior has one value and is then completed. The following list describes some of the properties of `Promises`:

- Guaranteed future (not always desirable in Web applications)
- Immutable
- Single value (not always desirable in Web applications)
- Caching
- Invoked immediately
- Cannot be canceled
- Cannot be reused

Hence, Promises are suitable when a future result is guaranteed and returns a single value. The following is a simple example of using a `Promise`:

```
getDataFromSomewhere(input)
    .then(data => {
        doSomethingHere(data);
```

```
        return getMoreData(data.id);
    })
    .then(data => {
        doSomethingHere(data);
        return getMoreData(data.id);
    })
```

In case you didn't already know, some of the operators that are available for `Observables` are also available as methods in ECMA5. Before we delve into FRP code samples, let's look at how to use some of those methods in the next section.

Using Operators without FRP

The following code block shows you how to chain the `filter()` and `map()` methods to process a set of integers:

```
var source = [0,1,2,3,4,5,6,7,8,9,10];

var result1 = source.map(x => x*x)
                    .filter(x => x % 5 == 0);
console.log("result1: "+result1);
// output=?

var result2 = source.filter(x => x % 5 == 0)
                    .map(x => x*x)
// output=?
```

Note that the output for `result1` and `result2` is the same. If possible, specify `filter()` methods before the `map()` methods to perform an "up-front" reduction in data (but see the caveat below). In the preceding example, the performance difference is probably undetectable, but if you change the source to include the first million positive integers, you will probably see a difference in performance.

As mentioned earlier, there is an important caveat regarding the order of operations: The `filter()` method and the `map()` method sometimes produce *different* results when they are invoked in the opposite order. For example, the following code block is a modified version of the preceding code block (modifications are shown in bold) that illustrates this point:

```
var source = [0,1,2,3,4,5,6,7,8,9,10];

var result3 = source.map(x => 2*x)
                    .filter(x => x % 4 == 0);
console.log("result3: "+result3);   // [0,4,8,12,16,20]
```

```
var result4 = source.filter(x => x % 4 == 0)
                    .map(x => 2*x) // [0,8,16]
```

The variable `result3` has the value `[0,4,8,12,16,20]`, whereas the variable `result4` has the value `[0,8,16]`.

An Analogy Regarding `Observables`

If you are new to `Observables`, or find yourself struggling with code samples that contain `Observables`, this section provides a humorous analogy by Venkat Subramanian with a clever insight into the world of `Observables`.

First, `Observables` involve some of the key concepts: chaining intermediate operators (such as `map()`, `filter()`, and so forth) and then (possibly later) invoking a terminal operator (such as `subscribe()` or `forEach`) in order to "make stuff happen..

Skipping the syntax-related details, consider the following pair of `Observables` in JavaScript that involve the intermediate operators `map()` and `filter()`:

```
var source = [0,1,2,3,4,5,6];

var result1 = source.map(x => 3*x)
                    .filter(x => x % 4 == 0);
console.log("result1: "+result1);

var result2 = source.map(x => 3*x)
                    .filter(x => x % 4 == 0)
                    .subscribe();
console.log("result2: "+result2);
```

Question: What is the difference between `result1` and `result2`?

Answer: Only `result2` contains the terminal operator `subscribe()`.

Result: The first `console.log()` displays nothing, and the second displays the numbers `0` and `12`.

Now let's read an entertaining (yet meaningful) analogy from Venkat Subramanian, who explains `Observables` by recounting a story of his wife and two teenaged sons, all of whom are watching television in their living room:

```
Mother: "It's time to switch off the TV".
Sons:    [No response.]
```

```
Mother: "It's time to take out the trash."
Sons:   [Nobody moves.]
Mother: "You need to start working on your homework."
Sons:   [Still nothing.]
Some time passes…
Mother: "I'm going to get your father."
Sons:   [Leaping into action…]
```

Hopefully you realize that the first three requests by the mother are similar to intermediate operators, and her final statement is analogous to a terminal operator, which starts the execution of the first three requests.

If this analogy has triggered a "lightbulb moment" for you regarding `Observables`, intermediate operators, and terminal operators, the good news is that many of the code samples in this chapter will be simpler for you to understand.

JavaScript Files for RxJS

The JavaScript files for `RxJS` are available via a content delivery network (CDN), and they consist of roughly 10 different files. The "core" JavaScript files that you need to include in an HTML Web page are shown here:

```
<script src="http://cdnjs.cloudflare.com/ajax/libs/rxjs/4.1.0/
                                                          rx.js">
</script>
<script src="http://cdnjs.cloudflare.com/ajax/libs/rxjs/4.1.0/
                                                    rx.async.js">
</script>
<script src="http://cdnjs.cloudflare.com/ajax/libs/rxjs/4.1.0/
                                                  rx.binding.js">
</script>
```

The version numbers may be different as this book goes to print. In addition, keep in mind that `RxJS` v5 is currently in beta. Additional `RxJS` files are available here (the first one is for animation effects):

```
<script src="http://cdnjs.cloudflare.com/ajax/libs/rxjs/4.1.0/
                                                     rx.time.js">
</script>
<script src="http://cdnjs.cloudflare.com/ajax/libs/rxjs/4.1.0/
                                              rx.coincidence.js">
</script>
<script src="http://cdnjs.cloudflare.com/ajax/libs/rxjs-
                                             dom/2.0.7/rx.dom.js">
```

```
</script>
```

With all of the preliminary details discussed, let's delve into various intermediate operators and terminal operators in the next section.

Intermediate and Terminal Operators

Think of `Observables` as streams or sets of data that can comprise any number of items (arbitrary time). `Observables` generate values when they are "subscribed," and can then be canceled and restarted (which is not the case for `Promises`). `Observables` help you avoid the "callback hell" that can occur in asynchronous code that does not use `Observables` or `Promises`.

Operators

Operators are methods in `Observables`. Operators can be intermediate or terminal (discussed later), and they allow you to compose new `Observables`. Common operators include `filter()`, `map()`, `reduce`, and `merge()`.

In the event that you need to create a custom operator in RxJS, you can do so with the following syntax:

```
Rx.Observable.prototype.myCustomOperator = // define
something here
```

You can use method chaining with operators with the following general syntax:

```
let obs = Rx.Observable
            .firstOperator()
            .secondOperator()
            .evenMoreOperatorsIfYouWant()
            .subscribe(....); // now stuff happens
```

Result: `obs` is an `Observable` that is "connected" to a source.

The `subscribe()` and `unsubscribe()` Operators

The `subscribe()` method must be invoked to generate data. By way of illustration, consider the following code block:

```
var source1 = Rx.Observable
                .range(0, 20)
                .filter(x => x < 4)

var source2 = Rx.Observable
                .range(0, 20)
                .filter(x => x < 4)
                .subscribe(x => console.log("x = "+x))
```

The first `Observable` does not generate output because there is no `subscribe()` method, whereas the second Observable displays the integers between `0` and `3` inclusive.

On the other hand, the `unsubscribe()` will "tear down" a producer, which means that the producer will stop producing data. The following code block shows you how to invoke the `unsubscribe()` method:

```
let x = Rx.Observable.(...)
let result = x.subscribe(...)
// do something here...
result.unsubscribe();
```

After invoking the `unsubscribe()` operator, you can restart an `Observable` and "resubscribe" to that `Observable`, which is not possible with `Promises`. (A proposal was submitted to the TC39 committee to add support for canceling `Promises` and was later withdrawn.)

The subscribe() and forEach() Operators

RxJS 4.0 follows the ES6 specification (but not the ES7 specification), in which `subscribe()` and `forEach()` are the same. However, RxJS 5.0 follows the ES7 specification, in which `subscribe()` and `forEach()` are different, as explained in the following paragraphs.

The syntax for the `subscribe()` method in RxJS 5.0 is shown here:

```
public subscribe(observerOrNext: Observer | Function, error:
Function, complete: Function): Subscription
```

`Observable.subscribe` returns a *subscription* token that enables you to cancel your subscription. This functionality is useful when the duration of the subscribed event is unknown, or if you need to perform an early termination.

The syntax for the `forEach()` method in RxJS 5.0 is shown here:

```
public forEach(next: Function, PromiseCtor?:
PromiseConstructor): Promise
```

`Observable.forEach` returns a *promise* that either resolves (or rejects) based on whether or not the `Observable` completes (or fails). Once again, remember that a `Promise` cannot be canceled.

Keep in mind that the vast majority of online code samples currently use RxJS 4.0, so it's probably worth your while to learn RxJS 4.0 as well as RxJS 5.0.

Converting Data Sources to `Observables`

You can "convert" other sources of data into an `Observable` with several methods, as shown here:

```
Observable.of(...)
Observable.from(promise/iterable/observable);
Observable.fromEvent(...)
```

The next set of subsections contains code samples that illustrate how to use variable operators in `RxJS`.

Using `range()` and `filter()` Operators

Listing 9.1 displays the contents of the `ObservableRangeFilter1.html` that illustrates how to define `Observables` using different syntax styles.

LISTING 9.1 `ObservableRangeFilter1.html`

```
<html>
 <head>
  <meta charset="utf-8">
  <title>Working with Observables</title>
 </head>

 <body>
  <script src="http://cdnjs.cloudflare.com/ajax/libs/
                                        rxjs/4.1.0/rx.js">
  </script>
  <script src="http://cdnjs.cloudflare.com/ajax/libs/
                                  rxjs/4.1.0/rx.async.js">
  </script>
  <script src="http://cdnjs.cloudflare.com/ajax/libs/
                                rxjs/4.1.0/rx.binding.js">
```

```
  </script>

  <script>
    var source1 = Rx.Observable
                    .range(0, 20)
                    .filter(x => x < 4)

    var source2 = Rx.Observable
                    .range(0, 20)
                    .filter(x => x < 4)
                    .subscribe(x => console.log("#2 x = "+x))

    var source3 = Rx.Observable
                    .range(1, 5)
                    .subscribe(
                      x  => console.log('onNext:  %s', x),
                      e  => console.log('onError: %s', e),
                      () => console.log('onCompleted'));
  </script>
 </body>
</html>
```

Listing 9.1 contains three `Observables`, the first of which does not generate any output because there is no `subscribe()` method. The second `Observable` filters the integers in the range `(0,20)` to those that are less than `4` and then displays their value via `console.log()`. The third observable iterates over the integers in the range `(1,5)` and displays their value via a `console.log()` method.

Launch the code in Listing 9.1 and you will see the following output:

```
#2 x = 0
#2 x = 1
#2 x = 2
#2 x = 3
onNext:  1
onNext:  2
onNext:  3
onNext:  4
onNext:  5
onCompleted
```

The next section contains an example that shows you how to chain the `from()` and `map()` intermediate operators.

Using `from()` and `map()` Operators

Listing 9.2 displays the contents of `ObservableMapUpper1.html` that illustrates how to use an `Observable` to "reactify" an HTML Web page.

LISTING 9.2 ObservableMapUpper1.html

```
<html>
 <head>
  <meta charset="utf-8">
  <title>Working with Observables</title>
 </head>

 <body>
  <script src="http://cdnjs.cloudflare.com/ajax/libs/
                                          rxjs/4.1.0/rx.js">
  </script>
  <script src="http://cdnjs.cloudflare.com/ajax/libs/
                                    rxjs/4.1.0/rx.async.js">
  </script>
  <script src="http://cdnjs.cloudflare.com/ajax/libs/
                                  rxjs/4.1.0/rx.binding.js">
  </script>

  <script>
    Rx.Observable.from(['a1','a2','a3'])
                 .map((item) => {
                    item = item.toUpperCase()+item;
                    return item;
                 })
                 .subscribe(str => console.log("item: "+str));
  </script>
 </body>
</html>
```

Listing 9.2 defines an `Observable` from an array of strings. The `map()` method converts each array item to its uppercase form and then appends the initial array item. The `subscribe()` method simply displays the value of each string that is created inside the `map()` method. Launch the code in Listing 9.2 in a browser and you will see the following output:

```
A1a1
A2a2
A3a3
```

The next section contains an example that shows you how to chain the `interval()`, `take()`, and `map()` intermediate operators.

Using the `interval()`, `take()`, and `map()` Operators

Listing 9.3 displays the contents of the `ObservableTake.html` that illustrates how to "interleave" the output from two `Observables` in an HTML Web page.

LISTING 9.3 ObservableTake.html

```
<html>
 <head>
  <meta charset="utf-8">
  <title>Working with Observables</title>
 </head>

 <body>
  <script src="http://cdnjs.cloudflare.com/ajax/libs/
                                    rxjs/4.1.0/rx.js">
  </script>
  <script src="http://cdnjs.cloudflare.com/ajax/libs/
                               rxjs/4.1.0/rx.async.js">
  </script>
  <script src="http://cdnjs.cloudflare.com/ajax/libs/
                             rxjs/4.1.0/rx.binding.js">
  </script>
  <script src="http://cdnjs.cloudflare.com/ajax/libs/
                                rxjs/4.1.0/rx.time.js">
  </script>

  <script>
    var source1 = Rx.Observable
                    .interval(1000)
                    .take(4)
                    .map(i => ['1','2','3','4','5'][i]);
    var result1 = source1.subscribe(x => console.log
                                           ("x = "+x));

    var source2 = Rx.Observable
                    .interval(500)
                    .take(4)
                    .map(i => ['1','2','3','4','5'][i]);

    var subscription = source2.subscribe(
      x  => console.log('source2 onNext:  %s', x),
      e  => console.log('source2 onError: %s', e),
      () => console.log('source2 onCompleted'));
  </script>
 </body>
</html>
```

Listing 9.3 defines two `Observables` that invoke the `interval()`, `take()`, and `map()` operators. The first `Observable` emits data every `1000` milliseconds whereas the second `Observable` emits data every `500` milliseconds. The first `Observable` invokes the `subscribe()` method that contains a `console.log()` statement for displaying data items. The second observable also invokes the `subscribe()` method with a different syntax: Three functions are specified that handle the `next`, `error`, and `completed` events, respectively.

Launch the code in Listing 9.3 and you will see the following output:

```
source2 onNext:  1
x = 1
source2 onNext:  2
source2 onNext:  3
x = 2
source2 onNext:  4
source2 onCompleted
x = 3
x = 4
```

At this point you have an understanding of how to combine some intermediate operators. RxJS supports many other operators, and the next section contains a high-level and rapid introduction to some of those operators.

Other Intermediate Operators

In addition to the `filter()` and `map()` operators that you saw earlier in this chapter, `Observables` support the following useful operators, most of which have intuitive names:

- reduce()
- first()
- last()
- skip()
- toArray()
- isEmpty()
- retry()
- startWith()

Even if you have not seen the preceding operators, you can probably surmise the result of this `Observable`:

```
var source = Rx.Observable
              .return(8)
              .startWith(1, 2, 3)
              .subscribe(x => console.log("x = "+x));
```

The `retry()` Operator

The `retry()` operator enables you to make multiple attempts to access data from an external website. Two examples of this syntax are shown here:

```
myObservable.retry(3);
myObservable.retryWhen(errors => errors.delay(3000));
```

The next section provides a list of some intermediate operators for merging and joining data streams in `Observables`.

A List of Merge/Join Operators

In Chapter 4 you learned about the `forkJoin()` intermediate operator that merges the data returned from HTTP requests from multiple endpoints. The following list contains various intermediate operators that perform merging or joining operations on streams of data:

- merge()
- mergeMap()
- concat()
- concatMap()
- switch()
- switchMap()
- zip()
- forkJoin()
- withLatestFrom()
- combineLatest()

Some of these are intuitively named (such as `concat()` for concatenating output), yet there are some subtle differences. For example, the `merge()` operator combines multiple `Observables` into one `Observable`, with the *possibility of interleaving* data values. On the other hand, `concat()` combines multiple `Observables` sequentially into one `Observable`, *without interleaving* any data.

Read the online documentation for the intermediate operators that interest you (or learn about all of them if you have the time!).

A List of Map-Related Operators

In Chapter 4 you saw several examples that use the `map()` intermediate operator, usually to convert an input stream into JavaScript Object Notation (JSON)-based data. A list of map-related operators is shown below:

- map()
- flatMap()
- flatMapLatest()
- mergeMap()
- concatMap()
- switchMap()
- flatten()

Recall from Chapter 4 how you used the map() operator to convert a string into JSON-based data. A more powerful operator is concatMap(), which uses concat() to ensure that intermediate results are not interleaved, and then the map() operator is applied.

As you learned in the previous section, intermediate results can be interleaved with the merge() operator. Because flatMap() uses the merge() operator, intermediate results can be interleaved with flatMap() as well (but not with the concatMap() operator).

Read the online documentation for the intermediate operators that interest you, and in particular, learn about the difference between the flatten() and flatMap() operators.

The timeout() Operator

The timeout() operator is useful for detecting if an Observable has not produced a value after a specified time period:

```
Rx.Observable
  .fromEvent(document, 'dragover')
  .throttle(350)
  .map(true)
  .timeout(1000, Rx.Observable.just(false))
  .distinctUntilChanged();
```

Keep in mind that timeout() will unsubscribe from the source Observable and subscribe with the Observable that was supplied as a parameter to timeout(); hence, you must resubscribe to that Observable to receive further results. The other point to remember is that the map() operation is placed after the throttle() operator to avoid unnecessary mapping operations of values that are not used.

Cold versus Hot Observables

A *cold* observable is comparable to watching a recorded movie (e.g., viewed in a browser). Although users navigate to the same URL at different times, all of them see the entire contents of the movie. In the case of cold observables, a new producer (movie instance) is created for each consumer (which is analogous to a person watching the movie).

By contrast, a *hot* observable is comparable to watching a live online presentation. Users navigate to a website at different times, and instead of seeing the entire presentation, they see only the portion from the point in time that they launched the presentation. In the case of cold observables, the same producer (streaming presentation) is used for each consumer (person watching the presentation).

Listing 9.4 displays the contents of the Web page `ColdObservables1.html` that illustrates a sequence of cold `Observables`, and how to convert them to hot observables.

LISTING 9.4 ColdObservables1.html

```
<html>
 <head>
  <meta charset="utf-8">
  <title>Working with Cold Observables</title>
 </head>

 <body>
  <script src="http://cdnjs.cloudflare.com/ajax/libs/
                                          rxjs/4.1.0/rx.js">
  </script>
  <script src="http://cdnjs.cloudflare.com/ajax/libs/
                                  rxjs/4.1.0/rx.binding.js">
  </script>

  <script>
    // this is a cold observable:
    let obs = Rx.Observable
                .create(observer => {
                   observer.next(13);
                });

    // first subscriber (displayed first):
    obs.subscribe(n => console.log("subscriber 1: " + n));

    // second subscriber (displayed second):
    var subscription = obs.subscribe(
      x  => console.log('Called next:  %s', x),
      e  => console.log('Called error: %s', e),
      () => console.log('Called completed')
    );
    subscription.dispose();

    // delay and add a subscriber (displayed fourth)
    setTimeout(() =>
      obs.subscribe(v => console.log("delayed: " + v)), 2000)

    // subscribe again (displayed third):
    obs.subscribe(n => console.log("subscriber 2: " + n));
```

```
  </script>
 </body>
</html>
```

Listing 9.4 contains a cold `Observable` followed by two `Subscribers` to that `Observable`.

Launch the code in Listing 9.4 and you will see the following output:

```
subscriber 1: 13
Called next:  13
subscriber 2: 13
delayed: 13
```

NOTE *All `Observables` are cold by default, whereas all `Promises` are hot by default.*

One way to convert a cold `Observable` into a hot `Observable` is via the `publish()` operator, as shown here:

> *http://blog.thoughtram.io/angular/2016/06/16/cold-vs-hot-observables.html*

Reactifying an HTML Web Page

Web pages that contain a button element for accessing external data are good candidates for the use of `Observables`.

Listing 9.5 displays the contents of `ObservableDivElement2.html` that illustrates how to use an `Observable` to reactify an HTML Web page.

LISTING 9.5 ObservableDivElement2.html

```
<html>
 <head>
  <meta charset="utf-8">
  <title>Working with Observables and div Elements</title>
 </head>

 <body>
  <script src="http://cdnjs.cloudflare.com/ajax/libs/
                                          rxjs/4.1.0/rx.js">
  </script>
  <script src="http://cdnjs.cloudflare.com/ajax/libs/
                                    rxjs/4.1.0/rx.async.js">
  </script>
```

```
  <script src="http://cdnjs.cloudflare.com/ajax/libs/
                                        rxjs/4.1.0/rx.time.js">
  </script>

  <div id="div1">This is a DIV element</div>
  <div id="div2">This is a DIV element</div>

  <script>
    let div1 = document.querySelector('#div1')
    let div2 = document.querySelector('#div2')

    var stream = Rx.Observable
                   .interval(500)
                   .take(10)
                   .map(x => x*x)
                   .subscribe(x => {
                      div1.innerHTML += x;
                      div2.innerHTML += x;
                    })
  </script>
 </body>
</html>
```

Listing 9.5 contains two `<div>` elements, followed by a `<script>` element that references both of them. The `<script>` element also defines an `Observable` that emits the first `10` integers, with a `500`-millisecond delay between data items.

The next part of the `Observable` invokes the `map()` operator to compute the square of each data item. The `subscribe()` method appends the newly computed number to the current contents of both `<div>` elements.

Launch the code in Listing 9.5 and you will see the following output in your browser session upon completion of the `Observable`:

This is a DIV element0149162536496481

This is a DIV element0149162536496481

RxJS and SVG Graphics/Animation

Listing 9.6 displays the contents of the Web page `SVG Observables 1 Anim1.html` that illustrates how to use `Observable` to generate and display SVG `<ellipse>` elements with an animation effect in an HTML Web page.

LISTING 9.6 SVGObservables1Anim1.html

```
<html>
 <head>
  <meta charset="utf-8">
  <title>Observables and SVG</title>
 </head>

 <body>
  <script src="http://cdnjs.cloudflare.com/ajax/libs/
                              rxjs/4.1.0/rx.js"></script>
  <script src="http://cdnjs.cloudflare.com/ajax/libs/
                        rxjs/4.1.0/rx.async.js"></script>
  <script src="http://cdnjs.cloudflare.com/ajax/libs/
                         rxjs/4.1.0/rx.time.js"></script>

  <script src="https://ajax.googleapis.com/ajax/
                       libs/jquery/1.11.1/jquery.min.js">
  </script>

  <div id="outer1">
    <svg width="600" height="300" id="svg1" />
  </div>

  <script>
    var XPoints = [], factor=30;
    var minorAxis = 20,  majorAxis = 40, strokeWidth = 1;
    var colors = ["#FF0000", "#FFFF00", "#0000FF"];
    var color  = colors[0], colorIndex = 0;
    var svgNS  = "http://www.w3.org/2000/svg";
    var svg    = document.querySelector("#svg1");
    var svgDocument;

    window.onload = function(evt){
      svgDocument = document.getElementById("svg1").
                                            ownerDocument;

      // generate some positions for the ellipses
      for(var i=2; i<10; i++) {
         XPoints.push(i*factor);
      }

      var delay = Rx.Observable.empty().delay(1000);
      var items = Rx.Observable.fromArray(XPoints)
                   .map(function (x) {
                     return Rx.Observable.return(x).
                                            concat(delay);
                   })
                   .concatAll();

      items.subscribe(x => {
```

```
            // create an SVG graphics element:
            var elem = svgDocument.createElementNS(svgNS,
                                                     "ellipse");
            elem.setAttribute("fill",
                              colors[(colorIndex++)%colors.
                                                     length]);
            elem.setAttribute("stroke",
                              colors[(1+colorIndex)%colors.
                                                     length]);
            elem.setAttribute("stroke-width", strokeWidth);
            elem.setAttribute("cx", x);
            elem.setAttribute("cy", x);
            elem.setAttribute("rx", majorAxis);
            elem.setAttribute("ry", minorAxis);
            // append the SVG element to the <svg> element:
            $("#svg1").append(elem);
        });
    }
  </script>
 </body>
</html>
```

Listing 9.6 contains a `<script>` element that initializes an assortment of JavaScript variables. The next portion of Listing 9.6 obtains a reference to a `<div>` element whose `id` attribute has the value `svg1`, followed by a simple loop that initializes the JavaScript array `XPoints`. After initializing `XPoints`, there is an interesting block of code that creates `Observable items` based on the `XPoints` array, and also a delay property that creates an animation effect (i.e., a `1`-second delay between rendering adjacent SVG `<ellipse>` elements).

The third portion of Listing 9.6 invokes the `subscribe()` method, which retrieves the data values in the populated array to calculate the position of a set of ellipses. An SVG ellipse is created via the method `createElementNS()`, and then its attributes are assigned via the `setAttribute()` method. The ellipse is fully populated and rendered, with a 1-second delay between consecutive ellipses, until a total of 10 ellipses have been rendered.

RxJS and Mouse Events in an HTML Web Page

Listing 9.7 displays the contents of `SVGObservables1MouseMove1.html` that illustrates how to handle mouse-related events. This code sample creates a follow-the-mouse effect that renders a different colored ellipse whenever users move their mouse.

LISTING 9.7 SVGObservables1MouseMove1.html

```
<html>
 <head>
  <meta charset="utf-8">
  <title>Observables, Mouse Events, and SVG</title>
 </head>

 <body>
  <script src="http://cdnjs.cloudflare.com/ajax/libs/
                                          rxjs/4.1.0/rx.js">
  </script>
  <script src="http://cdnjs.cloudflare.com/ajax/libs/
                                    rxjs/4.1.0/rx.async.js">
  </script>
  <script src="http://cdnjs.cloudflare.com/ajax/libs/
                              rxjs/4.1.0/rx.coincidence.js">
  </script>
  <script src="http://cdnjs.cloudflare.com/ajax/libs/
                                  rxjs/4.1.0/rx.binding.js">
  </script>
  <script src="http://cdnjs.cloudflare.com/ajax/libs/
                                     rxjs/4.1.0/rx.time.js">
  </script>
  <script src="http://cdnjs.cloudflare.com/ajax/libs/rxjs-
                                       dom/2.0.7/rx.dom.js">
  </script>
  <script src="https://ajax.googleapis.com/ajax/ libs/
                               jquery/1.11.1/jquery.min.js">
  </script>

  <div id="outer1">
    <svg width="600" height="300" id="svg1" />
  </div>

  <script>
    var minorAxis = 30,  majorAxis = 10, strokeWidth = 1;
    var colors = ["#FF0000","#FFFF00","#0000FF","#FFFFFF"];
    var color  = colors[0], colorIndex = 0;
    var svgNS  = "http://www.w3.org/2000/svg";
    var svg    = document.querySelector("#svg1");
    var svgDocument;

    window.onload = function(evt){
      svgDocument = document.getElementById("svg1").
                                              ownerDocument;

      // set up RxJS-related stuff...
      var mouseDownEvt =
               Rx.Observable.fromEvent(svg,"mousedown");
      var mouseUpEvt   =
```

```
            Rx.Observable.fromEvent(svg,"mouseup");
      var mouseMoveEvt =
            Rx.Observable.fromEvent(document,"mousemove");

      mouseDownEvt.map(function () {
         return mouseMoveEvt.takeUntil(mouseUpEvt);
      })
      .concatAll()
      .subscribe(function (e) {
         // create an SVG graphics element:
         var elem = svgDocument.createElementNS(svgNS,
                                                 "ellipse");

         elem.setAttribute("fill",
                           colors[(colorIndex++)%colors.
                                                 length]);
         elem.setAttribute("stroke",
                           colors[(1+colorIndex)%colors.
                                                 length]);
         elem.setAttribute("stroke-width", strokeWidth);

         elem.setAttribute("cx", e.x);
         elem.setAttribute("cy", e.y);
         elem.setAttribute("rx", majorAxis);
         elem.setAttribute("ry", minorAxis);

         // append the SVG element to the <svg> element:
         $("#svg1").append(elem);
      });

      // set initial click position
      mouseDownEvt.subscribe(function (e) {
       //offsetX = e.x - parseInt(svg.offsetLeft);
       //offsetY = e.y - parseInt(svg.offsetTop);
      });
    }
  </script>
 </body>
</html>
```

Listing 9.7 contains a `<script>` element that initializes some JavaScript variables. The next portion of Listing 9.7 defines the `Observables` to handle mouse down, mouse up, and mouse move events. The third portion of Listing 9.7 invokes the `subscribe()` method, which retrieves the data values in the populated array to calculate the position of a set of ellipses.

An SVG ellipse is created via the method `createElementNS()`, and then its attributes are assigned via the `setAttribute()` method. The code uses the value of the JavaScript variable `colorIndex` as an index into the `colors` array, as shown here:

```
elem.setAttribute("fill",
                  colors[(colorIndex++)%colors.length]);
elem.setAttribute("stroke",
                  colors[(1+colorIndex)%colors.length]);
```

After the other mandatory attributes are initialized, the newly created SVG `<ellipse>` is appended to the Document Object Model (DOM) via the jQuery "$" function.

Additional examples of handling mouse events with `RxJS` are available here:

- *http://jsfiddle.net/dinkleburg/ay8afp5f*
- *http://reactivex.io/learnrx/*
- *http://rxmarble.com*
- *http://cycle.js.org/basic-examples.html*

Other code samples involving `RxJS` and SVG graphics/animation are located here:

https://github.com/ocampesato/rxjs-svg-graphics

The following website illustrates how to create a toggle button with `RxJS`:

https://www.themarketingtechnologist.co/create-a-simple-toggle-button-with-rxjs-using-scan-and-startwith

An Observable Form

The code sample in this section shows you how to combine the intermediate operator `interval()` with the `async` operator in an Angular application to emit integers at regular intervals.

Copy the directory `ObservableForm` from the companion disc into a convenient location. Listing 9.9 displays the contents of `app.component.ts` that illustrates how to watch for changes in an `<input>` element and then dynamically update another `<input>` element with a randomly selected last name from an array of names.

LISTING 9.9 app.component.ts

```
import { Component }   from '@angular/core';
import { FormBuilder } from '@angular/forms';
```

```
import { FormGroup }   from '@angular/forms';
import 'rxjs/Rx'

@Component({
  selector: 'app-root',
  template: `
    <div>
      <h2>An Observable Form</h2>

      <form [formGroup]="myForm"
            (ngSubmit)="onSubmit(myForm.value)">

        <div class="field">
          <label for="fname">fname</label>
          <input type="text" id="fname"
                 [formControl]="myForm.controls['fname']">
        </div>

        <div class="field">
          <label for="lname">lname</label>
          <input type="text" id="lname"
                 [formControl]="myForm.controls['lname']">
        </div>

        <button type="submit">Submit</button>
      </form>

      <div class="field">
        <label for="randomName">Random:</label>
        <input id="randomName" [(ngModel)]="randomName">
      </div>
    </div>
  `
})
export class AppComponent {
  myForm: FormGroup;
  randomName = "";
  lNames = ["Anderson", "Smith", "Jones", "Edwards"];

  constructor(fb: FormBuilder) {
    this.myForm = fb.group({
      'fname': [''],
      'lname': ['']
    });

    this.myForm.controls['fname'].valueChanges
        .debounceTime(750)
        .subscribe(data => this.getUserName(data));
  }
```

```
  getUserName(fname) {
      var index = Math.floor(20*Math.random())+1;
      var lname = this.lNames[index%this.lNames.length];
      this.randomName = lname;
  }

  onSubmit(value: string): void {
    console.log('you submitted value:', value);
  }
}
```

Listing 9.9 is a copy of the directory `FormBuilder` from Chapter 5, with the new code sections shown in bold. The new functionality is straightforward: When users enter a text string in the first `<input>` element and then pause for 750 milliseconds, the `subscribe()` method in the constructor invokes the `getUserName()` method with the inputted string (which is not actually used in this example). Next, the `getUserName()` method randomly selects a last name from the `lNames` array, and displays the selected name in the third `<input>` element.

As you can see, this type of functionality is useful when you want to perform dynamic validation before users click the `<submit>` button.

Unsubscribing in Angular Applications

The Angular documentation indicates that Angular will invoke the unsubscribe method on your behalf for the Observables that you have defined in an Angular application. However, there is a bug (discovered as this book goes to print) in this functionality, which is described via a code sample in this blog post:

https://netbasal.com/when-to-unsubscribe-in-angular-d61c6b21bad3

An RxJS and Timer Example

The code sample in this section shows you how to combine the intermediate operator `interval()` with the `async` operator in an Angular application to emit integers at regular intervals.

Copy the directory `NGObservableTimer` from the companion disc into a convenient location. Listing 9.8 displays the contents of app.component.ts that illustrates how to use a simple timer in an Angular application.

LISTING 9.8 app.component.ts

```
import { Component } from '@angular/core';
import {Observable}  from 'rxjs/Observable';
import 'rxjs/add/observable/interval';

@Component({
  selector: 'app-root',
  template: '<h2>{{mytimer |async}}</h2>'
})
export class AppComponent {
   pause:number = 1500;
   mytimer = Observable.interval(this.pause);
}
```

Listing 9.8 contains a standard import statement, followed by two more import statements for an Observable and the interval() method. Next, a template property contains the Observable called mytimer whose output is "piped" to the async operator.

The AppComponent class contains the definition of mytimer that emits an integer every 1500 milliseconds (which is the value of pause in this example).

At this point you can easily modify this code sample by incorporating the Observables in this chapter.

RxJS Version 5

RxJS version 5 provides better debugging and better modularity. Version 5 also follows the ES7 specification, whereas RxJS version 4 follows the ES6 specification.

RxJS version 5 provides some notable performance improvement: RxJS version 5 is also 5 times faster (on average) than RxJS v4 in Google Chrome V8.

A code snippet with RxJS v5 in ES6:

```
import {Observable} from "rxjs/Observable";
import { map } from "rxjs/operator/map";
```

```
map.call(Observable.of(1,2,3), x => x*x)
                    .subscribe(console.log.bind(console));
```

A code snippet with RxJS v5 with Babel in ES6:

```
import {Observable} from "rxjs/Observable";
import { map } from "rxjs/operator/map";

Observable.of(1,2,3)::map(x => x*x)
          .subscribe(::console.log);
```

The following code snippet uses the syntax for RxJS version 5 in TypeScript:

```
import {Observable} from "rxjs/Observable";
import  "rxjs/add/operator/map";

Observable.of(1,2,3).map(x => x*x)
          .subscribe(console.log.bind(console));
```

Creating Observables in Version 5 of RxJS

The syntax for creating an Observable in version 5 is slightly different from the syntax in version 4: the "on" prefix has been dropped in version 5.

An example of the syntax for version 5 of RxJS is shown here:

```
let obs = new Observable(observer => {
    myAsyncMethod((err,value) => {
       if(err) {
          observer.error(err);
       } else {
          observer.next(value);  // older v4: onNext
          observer.complete();   // older v4: onComplete
        }
    });
});
```

Compare the preceding syntax with earlier examples of Observables in this chapter.

Caching Results in RxJS

Version 5 of RxJS supports the cache() operator, which enables you to cache results of an Observable. If you are using RxJS version 4, and you would like use caching in your code, the following link contains information about caching results:

http://www.syntaxsuccess.com/viewarticle/caching-with-rxjs-observables-in-angular-2.0

`d3.express`: The Integrated Discovery Environment

In Chapter 3 you saw some Angular applications that create D3-based graphics. As this book goes to print, Mike Bostock (creator of D3.js) is currently developing `d3.express`, which he describes here:

https://medium.com/@mbostock/a-better-way-to-code-2b1d2876a3a0

Based on the contents of this article, it's possible that `d3.express` will "play well" with RxJS (and hence its inclusion in this chapter), which bodes well for `d3.express`. This looks like yet another very interesting project from Mike Bostock that could provide more sophisticated data visualization functionality in Angular applications.

As this book goes to print, an alpha release of `d3.express` may be available, and you can sign up for early access here:

https://d3.express

Summary

This chapter introduced you to `FRP`, focusing primarily on `RxJS` for Web applications. You then learned about various operators in `FRP`, such as `filter()`, `map()`, and `reduce()`. You saw the similarities and differences between a `Promise` and an `Observable`. Then you learned about the difference between cold `Observables` and hot `Observables`, and how to convert a cold `Observable` into a hot `Observable`.

In addition, you learned how to reactify HTML elements in HTML Web pages. Finally, you saw how to generate SVG graphics and animation effects using `Observables`, and a follow-the-mouse example that displays SVG-based ellipses during mouse move events.

Finally, you saw how to combine the intermediate operator `interval()` with the `async` operator in an Angular application to emit integers at regular intervals.

CHAPTER 10

MISCELLANEOUS TOPICS

This chapter contains an eclectic mix of Angular features, including a brief description of two configuration files for Angular applications, ahead-of-time (AOT) compilation, the `ngc` compiler, the "tree-shaking" feature, the Webpack utility, and an assortment of other topics. These topics are covered lightly, in part because the details will probably change after this book has gone to print. However, you can search the relevant online documentation (or for blog posts) that contain the latest changes.

The first part of the chapter briefly discusses the configuration files `package.json` (required for the `npm` utility) and `tsconfig.json` (optional for the `tsc` utility). The second part of this chapter discusses AOT, its advantages, and two ways of invoking AOT.

The third part of this chapter discusses tree shaking and the `rollup` utility, as well as how to reduce the size of Angular applications. The fourth part of this chapter introduces you to Webpack, and how it's used with AOT. You will also learn about hot module replacement (HMR), which can shorten the development cycle, and some useful Angular utilities.

The final section delves briefly into deep learning via an Angular application that provides container-like functionality for the TensorFlow playground, where the latter provides an interactive visualization of highly customizable neural networks.

Angular 4.1.0

Angular version 4.1.0 was released as this book went to print. However, this version is a minor release, which means that there are no breaking changes. In addition, this version is a drop-in replacement for 4.x.x.

New Features

Version 4.1.0 adds full support for TypeScript 2.2 and 2.3; in fact, Angular is built with TypeScript 2.3. Fortunately, this change does not affect Angular 4.0, which was shipped with TypeScript 2.1.

In addition, Angular is now compliant with `StrictNullChecks` in TypeScript. Consequently, you can enable `StrictNullChecks` in Angular projects if you wish to do so.

The complete list of features and bug fixes for Angular 4.1.0 is available here:

https://github.com/angular/angular/blob/master/CHANGELOG.md

Angular Configuration Files

This section discusses the configuration files `package.json` (required for `npm`) and `tsconfig.json` (optional for `tsc`), which are part of every Angular application.

The `package.json` Configuration File

Listing 10.1 displays the JavaScript dependencies and their current version numbers in `package.json`, which are automatically generated by the `ng` command-line utility. Note that some version numbers may be slightly different by the time this book goes to print.

LISTING 10.1: package.json

```
{
  "name": "angular-example",
  "version": "1.0.0",
  "private": true,
  "description": "Example project.",
  "scripts": {
    "test:once": "karma start karma.conf.js --single-run",
    "build": "tsc -p src/",
    "serve": "lite-server -c=bs-config.json",
    "prestart": "npm run build",
    "start": "concurrently \"npm run build:watch\" \"npm run
                                                    serve\"",
    "pretest": "npm run build",
    "test": "concurrently \"npm run build:watch\" \"karma
                                start karma.conf.js\"",
    "pretest:once": "npm run build",
```

```
    "build:watch": "tsc -p src/ -w",
    "build:upgrade": "tsc",
    "serve:upgrade": "http-server",
    "build:aot": "ngc -p tsconfig-aot.json && rollup -c rollup-
                                                      config.js",
    "serve:aot": "lite-server -c bs-config.aot.json",
    "build:babel": "babel src -d src --extensions \".es6\"
                                                   --source-maps",
    "copy-dist-files": "node ./copy-dist-files.js",
    "i18n": "ng-xi18n",
    "lint": "tslint ./src/**/*.ts -t verbose"
  },
  "keywords": [],
  "author": "",
  "license": "MIT",
  "dependencies": {
    "@angular/common": "~4.0.0",
    "@angular/compiler": "~4.0.0",
    "@angular/compiler-cli": "~4.0.0",
    "@angular/core": "~4.0.0",
    "@angular/forms": "~4.0.0",
    "@angular/http": "~4.0.0",
    "@angular/platform-browser": "~4.0.0",
    "@angular/platform-browser-dynamic": "~4.0.0",
    "@angular/platform-server": "~4.0.0",
    "@angular/router": "~4.0.0",
    "@angular/tsc-wrapped": "~4.0.0",
    "@angular/upgrade": "~4.0.0",
    "angular-in-memory-web-api": "~0.3.1",
    "core-js": "^2.4.1",
    "rxjs": "5.0.1",
    "systemjs": "0.19.39",
    "zone.js": "^0.8.4"
  },
  "devDependencies": {
    "@angular/cli": "^1.0.0",
    "@types/angular": "^1.5.16",
    "@types/angular-animate": "^1.5.5",
    "@types/angular-cookies": "^1.4.2",
    "@types/angular-mocks": "^1.5.5",
    "@types/angular-resource": "^1.5.6",
    "@types/angular-route": "^1.3.2",
    "@types/angular-sanitize": "^1.3.3",
    "@types/jasmine": "2.5.36",
    "@types/node": "^6.0.45",
    "babel-cli": "^6.16.0",
    "babel-preset-angular2": "^0.0.2",
    "babel-preset-es2015": "^6.16.0",
    "canonical-path": "0.0.2",
    "concurrently": "^3.0.0",
    "http-server": "^0.9.0",
    "jasmine": "~2.4.1",
```

```
      "jasmine-core": "~2.4.1",
      "karma": "^1.3.0",
      "karma-chrome-launcher": "^2.0.0",
      "karma-cli": "^1.0.1",
      "karma-jasmine": "^1.0.2",
      "karma-jasmine-html-reporter": "^0.2.2",
      "karma-phantomjs-launcher": "^1.0.2",
      "lite-server": "^2.2.2",
      "lodash": "^4.16.2",
      "phantomjs-prebuilt": "^2.1.7",
      "protractor": "~4.0.14",
      "rollup": "^0.41.6",
      "rollup-plugin-commonjs": "^8.0.2",
      "rollup-plugin-node-resolve": "2.0.0",
      "rollup-plugin-uglify": "^1.0.1",
      "source-map-explorer": "^1.3.2",
      "tslint": "^3.15.1",
      "typescript": "~2.2.0"
    },
    "repository": {}
  }
```

The `scripts` section in Listing 10.1 specifies various commands that you can invoke from the command line. The next section in Listing 10.1 is a `dependencies` section that lists the modules that are required to compile and launch an application. The modules in this section are installed when you invoke `npm` from the command line. Notice that this section contains 12 Angular-specific modules that have version 4.0.0. The dependencies section is updated (always in alphabetical order) whenever you invoke `npm install` with the `--save` switch from the command line.

NOTE *The `devDependencies` section of `package.json` installs some executables on your machine that are referenced in the `scripts` section of `package.json`.*

If you want to use the `lite-server` executable as the server for this application, you can manually invoke `npm` to install `lite-server` in case it is not already installed on your machine.

The `devDependencies` section specifies modules that are only required for development, such as karma-related modules for performing tests.

The `tsconfig.json` Configuration File

The TypeScript compiler `tsc` uses the values of parameters in the `tsconfig.json` configuration file to transpile TypeScript files into

JavaScript files, instead of specifying parameter values from the command line.

Keep in mind that sometimes an Angular application does not work correctly, but no compilation errors are displayed in your browser's inspector. If this happens, invoke tsc from the command line to check for unreported errors.

Listing 10.2 displays the contents of tsconfig.json that contains configuration-related properties that will enable you to invoke tsc from the command line without specifying any arguments.

LISTING 10.2: tsconfig.json

```
{
  "compileOnSave": false,
  "compilerOptions": {
    "outDir": "./dist/out-tsc",
    "sourceMap": true,
    "declaration": false,
    "moduleResolution": "node",
    "emitDecoratorMetadata": true,
    "experimentalDecorators": true,
    "lib": [
      "es2016"
    ]
  }
}
```

Listing 10.2 contains various compiler-related properties and their values for the TypeScript compiler. Notice that the outDir property specifies the dist/out-tsc subdirectory as the location of generated files. The sourceMap is true, which means that a source map is generated during the transpilation process. The src subdirectory contains more configuration files whose contents you can peruse at your convenience. This concludes the brief section regarding configuration-related files in Angular applications.

What Is AOT?

Angular AOT is an acronym for ahead-of-time compilation, which involves compiling the application once (before the application is loaded in a browser), resulting in a faster load time.

As you have already seen, Angular compiles an application in a browser as it loads via just-in-time (JIT) compilation. However, JIT compilation incurs a runtime performance penalty, and the application is bigger because it includes the Angular compiler and a lot of unnecessary library code. JIT compilation can discover component–template binding errors at runtime, whereas AOT discovers template errors early and also improves performance via build-time compilation.

In addition, AOT is well-integrated with the Angular command-line interface (CLI). For example, the following command uses AOT compilation during the creation of a production build of an Angular application:

```
ng build --prod --aot
```

Advantages of AOT

Some of the advantages of AOT are listed below:

- Faster rendering
- Fewer asynchronous requests
- Smaller Angular framework download size
- Detect template errors earlier
- Better security
- Better performance
- Compile-time error reporting for templates
- Reduced application size
- Removal of dead code (tree shaking)

Until recently, mistakes in `ng` templates fail at runtime (sometimes silently), which made debugging Angular templates difficult. AOT will now report template errors at compile time, but currently AOT can only be used with Webpack 2. This situation might change at some point in the future.

AOT Configuration

There are two ways to enable AOT in Angular applications: use `@ngtools/webpack` (discussed below) or use the `ngc` utility (discussed in another section). The first approach provides more granular control, whereas the second approach is simpler and involves less configuration. Keep in mind an important point: Angular AOT will only work on code and metadata that is statically analyzable.

After reading the following sections you will be in a better position to decide which technique best suits your needs. You can refer to the following for additional information:

https://github.com/UltimateAngular/aot-loader

Setting up @ngtools/webpack

Install `@ngtools/webpack` and save it as a development dependency, as shown here:

```
npm install -D @ngtools/webpack
```

Next, add the following code to the configuration file `webpack.config.js` (Webpack is discussed later):

```
import {AotPlugin} from '@ngtools/webpack'

exports = { /* ... */
  module: {
    rules: [
      {
        test: /\.ts$/,
        loader: '@ngtools/webpack',
      }
    ] },
  plugins: [
    new AotPlugin({
      tsConfigPath: 'path/to/tsconfig.json',
      entryModule: 'path/to/app.module#AppModule'
    })
  ]
}
```

The `@ngtools/webpack` loader works with `AotPlugin` to enable `AoT` compilation. Note that the Angular CLI does not support custom configuration in every scenario.

Working with the ngc Compiler

The `ngc` compiler is a replacement for `tsc` and is configured in a similar fashion.

However, `ngc` attempts to inline cascading style sheets (CSS) without having the necessary context. For example, the `@import basscss-basic` statement in index.css results in an error because there is no indication that basscss-basic is located in node_modules.

On the other hand, @ngtools/webpack provides AotPlugin and a loader for Webpack that shares the context with other loaders/plugins. Consequently, when ngc is invoked by @ngtools/webpack, ngc can obtain information from other plugins (such as postcss-import) to compile things like @import 'basscss-basic'.

The tsconfig-aot.json File

The ngc utility relies on tsconfig-aot.json, which is a variant of tsconfig.json. The file tsconfig-aot.json contains AOT-oriented settings, an example of which is shown here:

```
{
  "compilerOptions": {
    "target": "es5",
    "module": "es2015",
    "moduleResolution": "node",
    "sourceMap": true,
    "emitDecoratorMetadata": true,
    "experimentalDecorators": true,
    "lib": ["es2015", "dom"],
    "noImplicitAny": true,
    "suppressImplicitAnyIndexErrors": true
  },

  "files": [
    "src/app/app.module.ts",
    "src/app/main.ts"
  ],

  "angularCompilerOptions": {
   "genDir": "aot",
   "skipMetadataEmit" : true
  }
}
```

Notice the two lines in bold that specify app.module.ts and main.ts, both of which are in the src/app subdirectory.

The Compilation Steps

Now, open a command shell and install these npm dependencies:

```
npm install @angular/compiler-cli @angular/platform-server -
                                                            -save
```

Next, invoke the `ngc` compiler in `node_modules/@angular/compiler-cli`, which creates AOT-related files via the following command:

```
./node_modules/.bin/ngc -p .
```

After the preceding command has completed, you will see an `aot` directory whose contents are shown here:

```
./aot
./aot/src
./aot/src/app
./aot/src/app/app.component.ngfactory.ts
./aot/src/app/app.component.ngsummary.json
./aot/src/app/app.module.ngfactory.ts
./aot/src/app/app.module.ngsummary.json
```

In brief, the component "factory" files create an instance of the component by combining the original class file and a JavaScript representation of the template in the component. Moreover, the generated factory references the original component class.

An example of AOT and dynamic Angular components is available here:

http://angularjs.blogspot.co.il/2017/01/understanding-aot-and-dynamic-components.html

Status of AOT, CLI, and Angular Universal

The Angular CLI does not support Angular Universal. A separate fork for a version of the Angular CLI supports Universal, but that fork does not support AOT. However, some of the key portions of Angular Universal will be placed in the Angular core.

Tree Shaking and the Rollup Utility

AOT compilation converts a greater portion of an Angular application to JavaScript. The next step invokes so-called tree shaking, which involves the removal of redundant code, thereby reducing the size of an Angular application. Keep in mind that tree shaking only works on JavaScript code.

Angular provides the tree-shaking utility called `rollup`, which performs a static code analysis to create a code bundle that excludes all exported code that is never imported. The `rollup` utility only works on ES2015 modules that contain both `import` and `export` statements.

Now install the `rollup` dependencies with this command:

```
npm install rollup rollup-plugin-node-resolve rollup-plugin-
commonjs rollup-plugin-uglify --save-dev
```

The `rollup-config.js` File

Create the configuration file `rollup-config.js` in the project root directory with the following contents:

```
import rollup      from 'rollup'
import nodeResolve from 'rollup-plugin-node-resolve'
import commonjs    from 'rollup-plugin-commonjs';
import uglify      from 'rollup-plugin-uglify'

export default {
  entry: 'src/app/main.js',
  dest: 'dist/build.js', // output a single application bundle
  sourceMap: false,
  format: 'iife',
  plugins: [
      nodeResolve({jsnext: true, module: true}),
      commonjs({
        include: 'node_modules/rxjs/**',
      }),
      uglify()
  ]
}
```

The `entry` attribute in the preceding file specifies `app/main.js` as the application entry point, and the `dest` attribute causes rollup to create the file `build.js` in the `dist` subdirectory.

Invoking the rollup Utility

Invoke the `rollup` utility from the command line by invoking the following command:

```
node_modules/.bin/rollup -c rollup-config.js
```

The preceding command creates the file `dist/build.js`, which you can reference in the HTML page `index.html`, as shown here:

```
<html>
  <head>
    <meta charset="utf-8">
    <title>FormBuilder</title>
    <base href="/">
    <meta name="viewport"
```

```
        content="width=device-width, initial-scale=1">
    <link rel="icon" type="image/x-icon" href="favicon.ico">
  </head>

  <body>
    <my-app>Loading...</my-app>
  </body>

  <script src="dist/build.js"></script>
</html>
```

Now launch the Angular application with this command:

```
npm run lite
```

After a few moments you will see a new browser session, and if everything worked correctly, you should see the Angular application.

Reducing the Size of Angular Applications

The use of Angular AOT, in conjunction with tree shaking, can reduce application code size and enable code to execute faster. The simplest option is to build a project for production with this command:

```
ng build --prod
```

However, you can reduce the file size even further with this command:

```
ng build --prod --aot
```

The preceding command removes unused code and the Angular compiler.

More information regarding the AOT compiler is available here:

https://angular.io/docs/ts/latest/cookbook/aot-compiler.html

Reducing the Size of Bundles

This section describes a simple process for measuring the bundles in an Angular application. The first step is to perform the following installation:

```
sudo npm install -g source-map-explorer
```

Next, build the application with the source maps, as shown here:

```
ng build --prod -sm
```

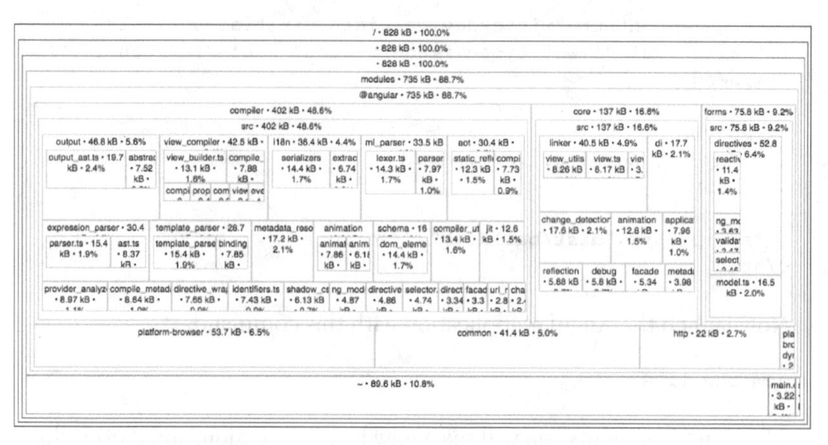

FIGURE 10.1 A map with the size of code in an Angular application.

Third, inspect one of the bundles, an example of which is available here (the name will be different for your application):

```
source-map-explorer dist/main.d357e5f7797d112767e6.bundle.js
```

The preceding command will launch a Chrome browser session and display the total code size, along with the percentage of that total that is attributable to various modules in the bundle.

Figure 10.1 displays the output from launching the preceding command, as displayed in a Chrome browser.

Angular Change Detection

Change detection in Angular is based on `zone.js`, which monitors all asynchronous events. Every component has its own change detector. Change detection (by default) checks if the value of any template expression has changed. Change direction is unidirectional and makes one pass, from top to bottom of component tree. Keep in mind that changes can only come from a component.

You can also programmatically specify the change detection strategy as shown in the following code snippet:

```
@Component({
    template: '...',
    changeDetection: ChangeDetectionStrategy.OnPush
})
export class MyComp {...}
```

The preceding code block contains the `changeDetection` property, whose value is `ChangeDetectionStrategy.OnPush`, which means that the component is re-rendered only when data in the component is modified.

As you already know, an Angular application is a hierarchical tree of Angular components, each of which has its own change detector. Whenever a component is modified, a change detection pass is triggered for the entire tree. In fact, Angular traverses the tree (from top to bottom) while scanning for changes.

The value `ChangeDetectionStrategy.Default` is the default value for `changeDetection` and `ChangeDetectionStrategy.OnPush` is another possible value. According to the Angular documentation (unfortunately, an explanation of "hydration" doesn't appear to be available in the documentation):

> OnPush means that the change detector's mode will be set to CheckOnce during hydration.
>
> Default means that the change detector's mode will be set to CheckAlways during hydration.

The preceding quote is from the following Angular documentation:

https://angular.io/docs/ts/latest/api/core/index/ChangeDetectionStrategy-enum.html#!#OnPush-anchor

However, if an Angular application uses immutable objects or Observables, it's possible to modify the change detection system to increase performance. Moreover, you can modify the behavior of the change detector of any component by specifying that checks are performed only during a change in one of its input values. Recall that an input value is an attribute that a component receives from elsewhere in the application.

Consider this code sample:

```
class Person {
   constructor(public name: string, public age: string) {}
}
@Component({
  selector: 'mycomp',
  template: '
    <div>
      <span class="name">{person.name}</span>
            lives in {person.city}.
    </div>
```

```
  `
})
class MyComp {
  @Input() person: Person;
}
```

To make change detection occur when a person changes (an input attribute), set its `changeDetection` attribute to `ChangeDetectionStrategy.OnPush`, as shown here:

```
import { Component, Input } from '@angular/core';
import { ChangeDetectionStrategy } from '@angular/core';

@Component({
  selector: 'my-selector',
  changeDetection: ChangeDetectionStrategy.OnPush,
  template: ' ....'
})
```

The zone.js Library

Angular uses the `zone.js` library for change detection in the following situations:

- When a Document Object Model (DOM) event occurs (such as click, change, and so forth)
- When an HTTP request is resolved
- When a timer is triggered (`setTimeout` or `setInterval`)

However, there are some situations where `zone.js` cannot detect changes, such as the following:

- Using a third-party library that runs asynchronously
- Immutable data
- Observables

What Is Webpack?

Although this chapter does not contain projects that rely on Webpack, it's the de facto utility for Angular applications, and its home page is located here:

https://webpack.github.io/

Webpack is a module bundler, which means that Webpack takes modules with dependencies and generates static assets representing those modules. Webpack is considered the latest "hotness" in Web application development, and in many ways Webpack supersedes the functionality of Gulp (but the latter is still relevant and useful). Fortunately, you can also combine Webpack with grunt, gulp, bower, and karma (see the online documentation for examples).

The goals of Webpack are as follows:

- Split the dependency tree into chunks loaded on demand.
- Keep initial loading time low.
- Every static asset should be able to be a module.
- Integrate third-party libraries as modules.
- Customize nearly every part of the module bundler.
- Suitability for big projects.

The mantra in Webpack is simple: Everything is a loader.

In addition, you can use AOT in conjunction with Webpack. An example of a Webpack configuration file for AOT is available here:

https://github.com/blacksonic/angular2-aot-webpack/blob/master/webpack.aot.config.js

Working with Webpack

The Webpack binary executable searches for a default configuration file called `webpack.config.js` is in the directory where you launch `webpack`. This configuration file contains an assortment of properties so that you do not need to specify them from the command line. Webpack supports many options, and you can see the entire list by invoking the following command:

```
webpack --help
```

Although version 2 of `webpack` was released in late 2016, you will still encounter tutorials that use version 1.x of `webpack`.

A Simple Example of Launching Webpack

Make sure you have already installed `npm` and then install Webpack with the following command:

```
npm install webpack -g
```

Next, install the Webpack development server with this command:

```
npm install webpack-dev-server -g
```

After completing the preceding steps, navigate to an empty directory and create two files:

```
entry.js
index.html
```

The contents of `entry.js` are here:

```
document.write("It works.");
```

The contents of `index.html` are here:

```
<html>
    <head>
        <meta charset="utf-8">
    </head>
    <body>
        <script type="text/javascript" src="bundle.js"
                                              charset="utf-8"></script>
    </body>
</html>
```

Now invoke the following command:

```
webpack ./entry.js bundle.js
```

After the preceding command has completed, you will find the file `bundle.js` in the same directory.

A Simple `webpack.config.js` File

Listing 10.3 displays the contents of `webpack.config.js` that is a sample Webpack configuration file.

LISTING 10.3: webpack.config.js

```
module.exports = {
    entry: "./entry.js",
    output: {
        path: __dirname,
        filename: "bundle.js"
    },
    module: {
        loaders: [
            { test: /\.css$/, loader: "style!css" }
        ]
    }
};
```

Listing 10.3 contains information about the name of the generated output file (in this case it's bundle.js) and a loader for CSS stylesheets.

Hot Module Reloading (HMR) and Webpack

Hot Module Reloading refers to dynamic recompilation of files. You can invoke HMR from the command line or specify HMR properties in `webpack.config.js`.

Specifically, there are three ways to invoke HMR from the command line.

The first option involves the `webpack-dev-server` utility (which you can install globally via `npm`), an example of which is shown here:

```
//Option #1: WDS is installed globally
webpack-dev-server --inline --hot
```

The second option involves installing `webpack-dev-server` as a dependency in `package.json`, as shown here:

```
//Option #2: WDS is installed as a dev-dependency
node_modules/webpack-dev-server/bin/webpack-dev-server.js
                                                --inline -hot
```

The third option involves specifying `webpack-dev-server` as one of the targets in the `scripts` element in `package.json`, as shown here:

```
//Option #3: modify scripts in package.json
{
 ...
  "scripts": {
    "start": "webpack-dev-server --inline --hot"
  }
  ...
}
```

Use whichever option best suits your needs.

AOT via a Modified `webpack.config.js`

In addition to the three ways of using AOT that were covered in the previous section, you can modify `webpack.config.js` to support AOT, as shown in Listing 10.4.

LISTING 10.4: webpack.config.js

```
var path = require('path');
var webpack = require('webpack');

module.exports = {
  devtool: 'eval',
  entry: [
    'webpack-dev-server/client?http://localhost:3000',
    'webpack/hot/only-dev-server',
    './src/index'
  ],
  output: {
    path: path.join(__dirname, 'dist'),
    filename: 'bundle.js',
    publicPath: '/static/'
  },
  plugins: [
    new webpack.HotModuleReplacementPlugin()
  ],
  module: {
    loaders: [{
      test: /\.js$/,
      loaders: ['react-hot', 'babel'],
      include: path.join(__dirname, 'src')
    }]
  }
};
```

Listing 10.4 expands the contents of Listing 10.3 by adding a `plugins` section and additional loaders in the `loaders` section. After making the preceding modifications to `webpack.config.js`, you can invoke HMR from the command line using one of the options described in an earlier section.

Because Angular 4.1.0 supports TypeScript 2.1 and above, please read the following caveat regarding AOT, TypeScript, and Webpack:

http://stackoverflow.com/questions/43276853/angular-4-aot-with-webpack/43282448

This concludes the section on AOT and the Webpack utility. The next section contains an Angular application that uses Angular Material.

Angular Material

Angular Material consists of Material Design components for Angular applications; its home page is located here:

https://github.com/angular/material2

Download and uncompress the zip file from the preceding link in a convenient location. Note the dependency on `HammerJS` in `package.json` for this code sample.

Navigate into the `material2-master` directory and install the dependencies with this command:

```
npm install
```

Next, launch the application with this command via `npm` (and not `ng`):

```
npm run demo-app
```

Navigate to the URL `localhost:4200` and you will see the output displayed in Figure 10.2.

Figure 10.2 displays examples of rendering user interface (UI) components using Angular Material (in a Chrome browser).

Click on the hamburger menu (top left corner of the screen) and you will see a list of various UI components, such as `Button`, `Card`, `Checkbox`, `Dialog`, `Grid List`, and `Menu`. Click any of these items and you will see examples of that UI component rendered with Angular Material.

You can look at the contents of `package.json`, which is almost two pages in length (so it won't be listed in this chapter).

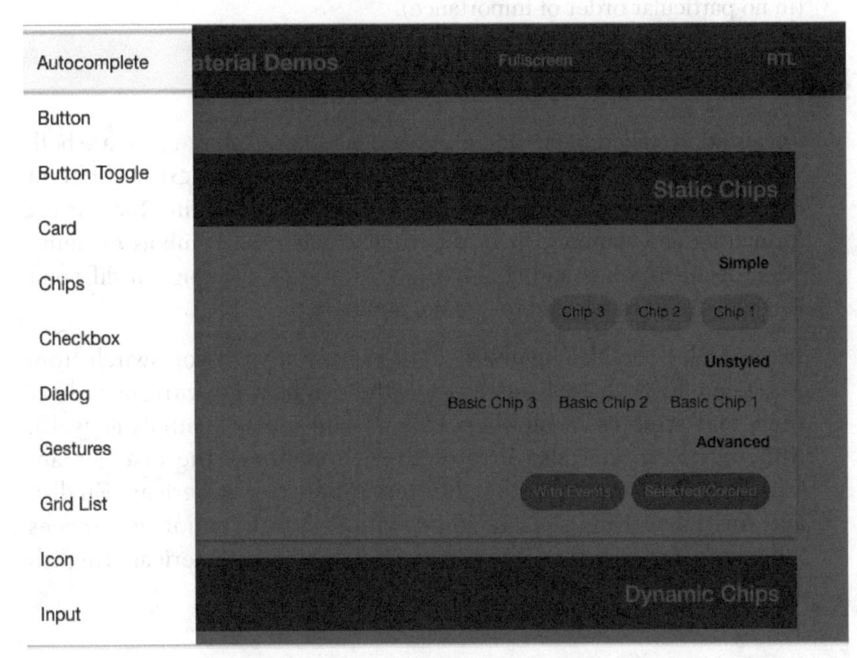

FIGURE 10.2 Some UI components with Angular Material.

The src/demo-app/demo-app subdirectory contains the code for the demo-app code sample, which contains more than 30 subdirectories, one for each of the UI components (including those that are listed above).

The preceding subdirectory also contains the TypeScript file demo-app-module.ts (also close to two pages in length), whose material-related import statement is shown here:

```
import {
  MaterialModule,
  OverlayContainer,
  FullscreenOverlayContainer,
  MdSelectionModule,
} from '@angular/material';
```

This code sample provides a good starting point to help you incorporate Material-based functionality in your own Angular applications.

Other Angular Functionality

There are several other topics that you can explore when you have the time to do so, and they are briefly mentioned in the following subsections (in no particular order of importance).

Support for I18n and L10n in Angular Applications

Localization (called L10n) and internationalization (called I18n) are both supported in Angular. Localization refers to displaying text in different languages, whereas internationalization refers to displaying the correct formatting and symbols, such as currency, date/time, symbols for numbers (decimal points and commas have different meanings in different languages), zip codes, and telephone numbers.

In general, L10n also involves I18n. For instance, if you switch from American English to French, then the symbols for currency, date/time, and symbols in numbers change (and other symbols as well). Although 118n can also involve L10n, sometimes the changes are "smaller." For example, the difference between American English and Australian English is regional, with relatively minor differences (obviously far less than the differences between American English and French).

The default language on an Android device (as well as laptops and desktops) depends on the country in which the device is used (e.g., American English in the United States and British English in the United Kingdom). Fortunately, users can easily change the default language via a menu option.

Working with a Component Container

Use `this.elementRef` to access a child component or create new elements. An example is shown here:

```
constructor(elem: ElementRef) {
   const tmp = document.createElement('div');
   const el = this.elementRef.nativeElement.cloneNode(true);

   // set the background color of elem
   this.elem.nativeElement.style.backgroundColor = 'blue';

   // use an Angular Renderer for portability
   this.renderer.setElementStyle(this.elem.nativeElement,
                                 'background-color',
                                 'blue');
}
```

Keep in mind that permitting direct access to the DOM is a security risk, as discussed in the following:

https://angular.io/docs/js/latest/api/core/index/ElementRef-class.html

The `ViewChild` Decorator

The `ViewChild` decorator enables you to access child elements in an Angular application, as shown here:

```
@ViewChild('input') input: ElementRef;
```

Angular supports the `@ViewChild` decorator to search the template for an element whose name is `input`. Notice that the type is `ElementRef` because a specific class name is not available.

A variant of the preceding code, when only one element exists, is shown here:

```
@ViewChild(ElementClassName) variableName: ElementClassName;
```

You also need to import `ElementClassName` in your Angular application.

Where to Specify a Service

There are several ways that you can specify a service, depending on what you intend to do with that service, as briefly discussed below.

- Option 1: Specify a service in `providers` in `NgModule` if you want a single instance of the service to be used/shared throughout the application.
- Option 2: Inject a service in the `providers` of a `Component` if you want every instance of the class to use the same instance of the service.
- Option 3: Inject a service in the constructor of a class if you want every instance of that class to have a different instance of the service.
- Option 4: Inject a service in the `viewProviders` of a `Component` if you want one instance per component and shared only with the component's *view* children, but not with the component's *content* children.

Consult the online documentation for more detailed information regarding the preceding scenarios.

Testing Angular Applications

This is an important topic that is not covered in this book. However, the following links provide excellent information (including code samples):

https://medium.com/google-developer-experts/angular-2-testing-guide-a485b6cb1ef0#.s7grqvua2

http://www.discoversdk.com/blog/writing-unit-tests-in-angular-2

https://blog.nrwl.io/essential-angular-testing-192315f8be9b#.yybrldt04

Useful Angular Utilities

There are various third-party utilities available that provide additional features, as a Chrome extension, as an installable module (via npm), or as a command-line executable. Two utilities that are briefly discussed below are the Augury Chrome extension, which that provides debugging support, and the `ngd` utility for displaying a dependency tree of components in Angular applications.

The Augury Chrome Extension

The Augury Chrome Developer Tools extension provides very good debugging support for Angular applications, and it can be downloaded from this link:

https://augury.angular.io/

The `augury` tab displays two panels where you can decide to view either the component tree or the router tree (if there is one) of an Angular component in the left panel. When you select a component in the left panel, the details of that component (including variables) are displayed in the right panel. In addition, you can dynamically modify the values of variables in the right panel and then apply those modifications in the current Angular application.

Figure 10.3 displays the sample contents of the `augury` console tab for a simple Angular application in a Chrome browser.

Figure 10.4 displays the sample contents of the `augury` Component Tree tab for a simple Angular application in a Chrome browser.

Displaying a Dependency Tree of Angular Components

The `ngd` utility is an open source utility that provides a hierarchical display of the components in an Angular application, and its home page is located here:

```
https://github.com/compodoc/ngd
```

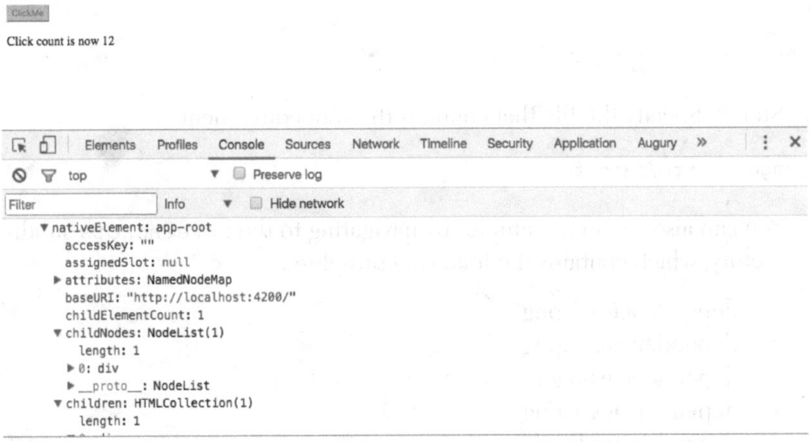

FIGURE 10.3 The Augury Console tab for an Angular application.

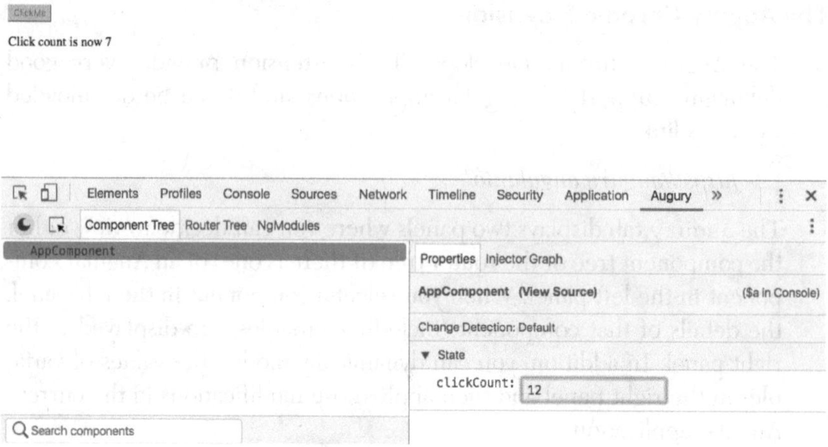

FIGURE 10.4 The Augury Component Tree tab for an Angular application.

Install ngd as follows:

```
npm install -g @compodoc/ngd-cli OR
yarn global add @compodoc/ngd-cl
```

Next, include some additional `ngd`-related code in an Angular application (as described in the `README.txt` file) to use the `ngd` utility. After adding the required code, navigate to the root directory of your Angular application and perform the following steps.

Step 1: Launch `ngd` as follows:

```
ngd OR
ngd -p ./tsconfig.json
```

Step 2: Specify the file that contains the root component:

```
ngd -f src/main.ts
```

You can also see some samples by navigating to the `screenshots` subdirectory, which contains the following samples:

- dependencies-1.png
- dependencies-2.png
- dependencies-3.gif
- dependencies-4.png
- dependencies.html
- dependencies.material2.svg

- dependencies.ng-bootstrap.svg
- dependencies.soundcloud-ngrx.svg

Compodoc

Compodoc is a documentation tool that generates static documentation of Angular applications, and its home page is located here (which includes a live demo):

https://github.com/compodoc/compodoc

Compodoc automatically generates a table of contents and provides various other features, such as themes, search capability, JSDoc light support, and it's also Angular CLI–friendly.

The website for the official documentation is located here:

https://compodoc.github.io/website/guides/getting-started.html

Angular and Deep Learning

This section contains the Angular application `AngularDLPG`, which is based on the code in the following "deep playground" GitHub repository. This repository provides neural networks based on Deep Learning:

https://github.com/tensorflow/playground

Deep playground is an interactive visualization of neural networks that is written in TypeScript using D3.js. In essence, the `AngularDLPG` application acts as a container for this interactive visualization.

The `AngularDLPG` application was created in three steps, starting with the `ng` utility, to create the baseline application. Next, the files `package.json` and `index.html` from the preceding GitHub repository were merged into the corresponding files in the `AngularDLPG` application. Third, the TypeScript files in the `src` subdirectory in the GitHub repository were copied into the `src` subdirectory of the `AngularDLPG` application.

Copy the `AngularDLPG` directory from the companion disc to a convenient location. This application contains the following TypeScript files (copied from the GitHub repository) in the `src` subdirectory:

- dataset.ts
- heatmap.ts
- linechart.ts

- main.ts
- nn.ts
- playground.ts
- polyfills.ts
- seedrandom.d.ts
- state.ts
- typings.d.ts

The preceding TypeScript files (excluding `typings.d.ts`) were also modified by the addition of the following code snippet:

```
import * as d3 from 'd3';
```

The preceding code snippet enables the TypeScript files to access the D3-related code, which is in the `node_modules` subdirectory.

Launch the Angular application by navigating to the `src` subdirectory and executing the following command:

```
ng serve
```

Figure 10.5 displays the contents of the TensorFlow playground in an Angular application in a Chrome browser.

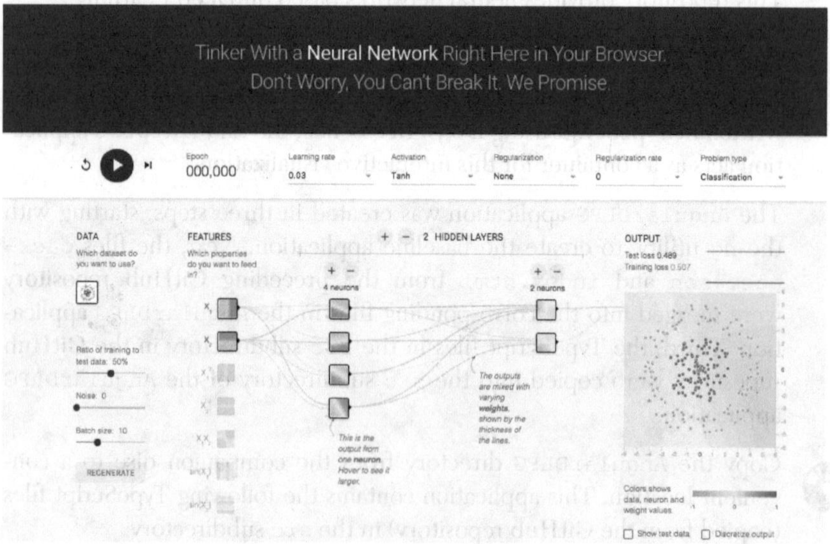

FIGURE 10.5 The TensorFlow playground in an Angular application.

Summary

This chapter started with an update regarding Angular 4.1.0, followed by an introduction to configuration files for the `npm` utility and the `tsc` utility. Next, you learned about AOT, which is a sophisticated part of Angular for optimizing the size and performance of Angular Web applications.

You also saw how to use the `ngc` compiler, the purpose of the tree-shaking feature and the `rollup` utility. Next, you learned about Angular change detection and the Webpack utility. You also learned about HMR (hot module reloading) in conjunction with Webpack. Then you saw an example of using Angular Material, which can enhance the aesthetic value of Angular applications.

You also learned about an assortment of other topics in Angular, such as support for `I18n`, the Augury Chrome Developer Tools extension for debugging, and the `ngd` utility for displaying the dependency tree of an Angular application.

Finally, you saw an Angular application that contains an interactive visualization of Deep Learning neural networks.

INDEX

D

E

F

G

H

I